Group Rationality in Scientific Research

Under what conditions is a group of scientists rational? How would rational scientists collectively agree to make their group more effective? What sorts of negotiations would occur among them and under what conditions? What effect would their final agreement have on science and society? These questions have been central to the philosophy of science for the last two decades. In this book, Husain Sarkar proposes answers to them by building on classical solutions – the skeptical view, two versions of the subjectivist view, the objectivist view, and the view of Hilary Putnam. Although he finds none of these solutions completely adequate, Sarkar retrieves what is of value from them, and also expropriates the arguments of John Rawls and Amartya Sen, in order to weave a richer, deeper, and more adequate theory of group rationality.

Husain Sarkar is professor of philosophy at Louisiana State University. A recipient of LSU's Distinguished Faculty Research Award, he is the author of *Descartes' Cogito: Saved from the Great Shipwreck.*

Group Rationality in Scientific Research

HUSAIN SARKAR
Louisiana State University

CAMBRIDGE UNIVERSITY PRESS
Cambridge, New York, Melbourne, Madrid, Cape Town,
Singapore, São Paulo, Delhi, Tokyo, Mexico City

Cambridge University Press
32 Avenue of the Americas, New York, NY 10013-2473, USA

www.cambridge.org
Information on this title: www.cambridge.org/9780521317986

First published 2007
First paperback edition 2011

A catalog record for this publication is available from the British Library

Library of Congress Cataloging in Publication data

Sarkar, Husain, 1942–
Group rationality in scientific research / Husain Sarkar.
p. cm.
Includes bibliographical references and index.
ISBN-13: 978-0-521-87113-6 (hardback)
ISBN-10: 0-521-87113-1 (hardback)
1. Science – Philosophy. 2. Reasoning. 3. Group decision making.
4. Group problem solving. 5. Research – Psychological aspect. I. Title.
Q175.32.R45S37 2007
507.2 – dc22 2006022359

ISBN 978-0-521-87113-6 Hardback
ISBN 978-0-521-31798-6 Paperback

For

My Mother and Father

Alas, I am so *unforgivably late:*

وَ وَصَّيْنَا الْإِنسَانَ بِوَالِدَيْهِ إِحْسَانًا

. . . . حَتَّىٰ إِذَا بَلَغَ أَشُدَّهُ وَبَلَغَ أَرْبَعِينَ

سَنَةً قَالَ رَبِّ أَوْزِعْنِي أَنْ أَشْكُرَ نِعْمَتَكَ الَّتِي

أَنْعَمْتَ عَلَيَّ وَعَلَىٰ وَالِدَيَّ

We have enjoined man to show kindness to his parents.... When he grows to manhood and attains his fortieth year, let him say: 'Inspire me, Lord, to give thanks for the favors You have bestowed on me and on my parents.'

The Koran, *Chapter 46, "The Sand Dunes," Verse 15*

Contents

Preface

The problem of group rationality can be fairly simply stated: Under what conditions is a group of scientists rational? There are a few scattered answers, some not so well known, some not so well reasoned. Our task is to examine some of these theories of group rationality to show why that problem is a marvelous puzzle, why that puzzle is yet unsolved, and why it needs solving. Hitherto, philosophers of science have dealt with the following problems: Under what conditions is a theory scientific? (Popper: when the theory is falsifiable.) Under what conditions is a theory making scientific progress? (Lakatos: when the theory is a successful research program.) Under what conditions is a decision to accept a scientific theory rational? (Bayes: when the prior probability of the scientific theory multiplied by the likelihood of the theory, the total divided by the probability of the evidence, is high.) Finally, under what conditions is a scientist generally rational; in other words, what is individual rationality? There is a paucity of theories on this; perhaps attempts at solving the problem of group rationality will stimulate interest in this area by making it evident that this last question lies intriguingly beneath the topsoil.

In 1983, the problem of group rationality was a new problem. By now, more than twenty years later, several philosophers have been engaged in the task of solving it; but in solving it, they have sometimes assumed that it is a problem of a different stripe. The problem of group rationality, I shall argue in Chapters 2 and 3, is a unique problem. In the first chapter, as in the last, the work of John Rawls plays a preeminent role, both when I appropriate his views to show how in some places the problem of group rationality is immensely enriched if we heed his work, as well as when I show why in other places we might do well to explore on our own.

In Chapter 2, I shall argue that it is *not* a problem to be solved by the strategy of using evolutionary mechanisms; it is *not* akin to the problems treated in game theory (with its fruitful and fascinating analysis and extension of the Prisoner's Dilemma, iterated or otherwise); and it is *not* an adapted version of the problem of social justice. (How applications of such approaches in the domain of group rationality can lead to unmitigated disaster is illustrated in Chapter 5.) I have also drawn upon the ingenious, marvelously inventive work of Amartya Kumar Sen in Chapter 3; that merest sketch should indicate to the reader the rich possibilities that Sen's approach in welfare economics and social choice theory, duly adopted, promises to the field of group rationality. But I shall also show in this chapter that the problem of group rationality is *not* simply another version of a problem in welfare economics.

Game theory, evolutionary dynamics, and welfare economics may eventually throw a good deal of light on the problem of group rationality; for now, however, we must perforce use the less formal, more substantive, traditional philosophical method – not a less likely route to harvesting rich results. Indeed, unless concepts custom designed for a theory of group rationality are available – as are concepts unique to the domain of justice – game theory, welfare economics and social choice theory, and evolutionary dynamics will not know *what* they are supposed to explain in the first place.

The purported solutions to the problem of group rationality have entered a phase that might be called classical orthodoxy. This book is about some of those attempts to solve that problem. I have allowed myself to reconstruct solutions, weaving materials extracted from the works of philosophers who have collectively, and masterfully, defined the field of methodology. Thus, although Paul Feyerabend, Thomas Kuhn, Imre Lakatos, and Karl Popper have never spoken of the problem of group rationality – not in any direct way, at any rate – I have scavenged their works to reconstruct various solutions to the problem. This task is performed in Chapters 4 through 7. This results in the presentation of the skeptical view, two versions of the subjectivist view, and the objectivist view. In brief, the skeptical view is presented as offering an epistemic challenge to anyone offering a theory of group rationality (on pain of accepting its own crazy view), while each successive view can be seen as claiming that it answers the skeptical challenge better, or more effectively, than do the preceding views. The arguments presented in the book do not move in a linear direction; there is some looping back, as when the skeptical view is recalled to plumb the depths of other views.

The penultimate chapter centers on aspects of the later philosophy of Hilary Putnam, maker and keeper of philosophical traditions. Putnam, as far as I can tell, has no interest in the problem of group rationality. But Putnam's iconoclasm is profitably used in this chapter – used, I say; not misused, I trust – to cast the notions of science, rationality, and relativism in a more revealing light, thanks in large measure also to Charles Sanders Peirce. What will emerge, I hope, is that the notion of individual rationality will appear utterly indispensable to solving the problem of group rationality. Moreover, if we are not careful with the notion of individual rationality, not only will the problem of group rationality remain unsolvable at the deepest level, we might also find ourselves sliding into relativism. I shall also show, vis-à-vis what I call the Williams problem, after Bernard Williams, that once we distinguish between a Social Utopia and a Scientific Utopia, we may no longer be able to claim that even an ideal democracy (Social Utopia) solves, in any significant way, the problem of group rationality; at most, democracy may be a necessary condition.

The final chapter records nine problems that lie at the heart of the investigation into the problem of group rationality; and, in so doing, it does not merely sketch these problems and the unique manner in which they are knotted together, it also signals a fascinating additional problem or two – none of them even touched upon in the rest of the book – namely, *why* do scientists owe allegiance to fellow scientists? And, given that they owe it, *what* will sustain that allegiance? Perhaps the answers to these questions ultimately lie in our speculation over a problem that lies considerably underground, namely, what is science *for*?

I have endeavored to show how deep the problem of group rationality is; why the classical theories fail to solve it; why new foundations are needed; what problems will need to be addressed in order to arrive at a more plausible solution; and, finally and above all, why the importance of the problem of group rationality – let alone its beauty – overshadows the problems that have occupied us these past fifty years or so in philosophy of science, thereby showing these problems their rightful place in the scheme of methodology. Despite their depth and scope, as is evidenced by the reconstruction, I argue that on the whole the vices of these classical theories of group rationality exceed their many virtues; hence, one must also engage in the task of dismantling.

As I view things, there is room for one more vision of group rationality, a vision that for now is buried in footnotes, or tracked only in the implications, several of them far-flung, of what is said in these pages. That vision will have to wait its turn for full expression in the second half of this project. For now, our task is to calculate the value of what we have on our hands; and calculate we must. The stakes are high, if I am right. For not only is this task about an ideal scientific group, it is also about reasons generally (and how they are anchored) and about utopias (and why we owe them allegiance). It is about what a society stands to reap if a rational scientific group flourishes in its midst – or what it stands to fear if it does not. I leave all that for now (since the rest of the book is occupied with it) and turn to acknowledging the debts I have incurred.

Acknowledgments

From what follows, it would be fair and reasonable to surmise that my debts – I record them in chronological order – are several, considerable, and, in a case or two, unpayable.

Paul Feyerabend discussed with me an earlier paper of mine, "Against *Against Method*; or, Consolations for the Rationalist," which now plays its own small part as section IV of Chapter 4, which centers upon Feyerabend's skeptical views. Of course, he found in that paper much with which to quarrel (an a priori truth). When I was a mere graduate student and he was at the height of his fame, he corresponded with me (and this continued for several years thereafter), giving me advice mingled with encouragement. I was back then too much of a Popperian, and he too anti-Popperian, for either of us to heed what the other said (old loves don't entirely wither away), and yet he was kind enough to write letters on my behalf. From an utterly chance encounter in 1979 (en route to Ian Hacking's National Endowment for the Humanities Summer Seminar at Stanford University) on an uphill street in Berkeley – we stood on the sidewalk and talked for well over an hour – I caught a glimpse of the man's core, especially when he talked of his loss in Imre Lakatos, that I had not known from our correspondence. Undoubtedly, he would have returned, as was his custom, with a scathing argument – or, more likely, laughed – at what I have written here; but not even *he* would have doubted that it was written with much affection, gratitude, and respect.

In the fall semester of 1983, Princeton University elected me its Visiting Fellow. Frequent lone and leisurely walks on and around the campus produced lots of first and fruitful ideas; I hold the incredible fall season of that year as my witness. The Institute for Advanced Study at Princeton, in the person of Donald McCloskey, invited me that December to give a talk on the problem of group rationality. In 1984, the National Endowment for the Humanities offered me a Summer Seminar Award, which enabled me to go to Harvard University. The seminar was led by Everett Mendelsohn on the topic "The Social Context of Modern Science." I worked on the problem of group rationality on the side when I could, and there I first crafted a crude sketch of the objectivist view, now Chapter 7.

Nearly everything by way of a first draft – no part of it was subsequently left untouched – was written between the two Octobers of 1984 and 1987. After a three-month recess, I revised the manuscript for approximately six months. I made further revisions in the summer of 1990 and the spring of 1991. Louisiana State University's Council on Research granted me a summer research award in 1985 and again in 1992, without which much of my work would not have been accomplished. Then, inexplicably – but not unhappily – for nearly a decade I abandoned the project as I turned to study the history of philosophy.

But, even during this interval, together with a couple of papers on scientific realism, I published earlier versions of Chapters 5 and 6 on the two types of the subjectivist view of group rationality, and earlier still a small part (section IV) of Chapter 4 on the view of the skeptic. I am grateful to the editors and publishers of *Philosophical Topics* and *Studies in History and Philosophy of Science* for permission to print here, duly revised, material that was originally published in their journals. Once again, I owe thanks to my son, Casim Ali, for the several diagrams in this book and for providing an engineer's help with Chapters 2 and 5. I added, deleted, and rewrote large parts of the manuscript for a year and a half beginning in late October 2003. A sabbatical leave for fall 2004 was crucial for much of my labor on the book; I am quite indebted to my university. Finally, the first six months of 2005, as well as of 2006, were spent reworking and restructuring, and thus was this work finalized. My gratitude to Russell Hahn for serving as production editor and, once again, as copy editor: no book should be without so patient and skillful a copy editor.

Let me here interpose an acknowledgment that has nothing to do with the making of this book (except indirectly, as a consequence of a long-ago event), and yet it is an acknowledgment I perforce must make, with due deference and publicly. Nearly a quarter of a century ago (and the reader, I hope, will understand why I waited so long to make this public), three philosophers – Baruch A. Brody, Samuel Gorovitz, and Leonard Linsky – and a chemist, Sean McGlynn, then Boyd Professor and Vice Chancellor for Research at Louisiana State University, acted in concert, judiciously and vigorously, to ensure the survival of my academic life, and ensure it they did. I am sorely in need of instruction on how to repay this debt.

This book is one half of the project; the other half is to propose an alternative to the theories of group rationality evaluated herein. Late in 2002, I sent a draft of the entire project to a few philosophers who were willing to aid in my cause. These were Joseph Agassi, Alexander Reuger, and Catherine Wilson. I have shaped the contents of the book, and remeasured its tone, in light of their advice and admonitions. Of the anonymous referees for Cambridge University Press, I thank two: for showing me how to illumine what was dark and obscure in the book, to tighten what was loose, and to provide an aerial view of the labyrinth. In that last advice I heard Seamus Heaney's voice: "So that the figure of the universe / And 'not just single things' would meet his sight." What has resulted is not faultless, but it *is* less faulty. I therefore proportionately thank them (concealing, for honor's sake, the extent of that proportion from the public's eye). When I was racked by uncertainty, Beatrice Rehl, my editor, admonished me thus: "Write the book you want to write," and then stood sentry over it. Those words have sustained me through the final rewriting. Finally, and once again, I owe a debt to Catherine Wilson: a lasting, burgeoning debt.

May 19, 2006
Baton Rouge, Louisiana

1

The Overview

Charles Robert Darwin (1809–1882) published *Origin of Species*[1] on November 24, 1859; by the day's end, all the printed copies – all 1,500 of them – had sold out.[2] The book was read avidly even by the laity – 500 copies went to Mudie's Circulating Library – and the revolution it initiated was off and running.[3]

[1] The full title of the book as it appeared in the first edition was *On the Origin of Species By Means of Natural Selection, or the Preservation of Favoured Races in the Struggle for Life.* The details of the references are given in the bibliography.

[2] Even that is not exactly true. The number of copies printed was 1,250, although Murray, Darwin's publisher, took orders for 1,500 copies. When Murray asked Darwin to send in the corrections post haste for a second edition, Darwin was most pleased. It is worth quoting what Darwin's wife, Emma Darwin, wrote to their son, William: "It is a wonderful thing the whole edition selling off at once & Mudie taking 500 copies. Your father says he shall never think small beer of himself again & that candidly he does think it very well written." Nine days later, on December 3, 1859, Mudie's Circulating Library advertised that *Origin of Species* was available to be borrowed. For a detailed account of this, and of the whirlwind that followed in the wake of Darwin's book, see Janet Browne's splendid biography, *Charles Darwin: The Power of Place,* especially Chapters 3 and 4.

[3] One would commit an egregious sin of omission if one did not mention the self-effacing, not-so-well-connected (in fact, ostracized by the community of his peers) co-discoverer of evolution, Alfred Russel Wallace. For a fascinating account of Wallace's own independent discovery of the theory of evolution, based on his researches in the Amazon (from 1848 to 1852) and the Malay Archipelago (1854 to 1862), and a judicious treatment of the evidence on the issue of priority, see Michael Shermer, *In Darwin's Shadow: The Life and Science of Alfred Russel Wallace,* especially Chapters 2 through 5.

The cardinal tenet of *Origin of Species* was that there is evolution in the biological world that can be explained by the principle of natural selection. Darwin had argued that biological species evolve (were not separately created) through competition for scarce resources, and that the winner in said competition is defined by differential reproduction (one who is able to leave behind more offspring than others). Among the key Darwinian ideas, in part inspired by Thomas Robert Malthus's 1798 *An Essay on the Principle of Population*,[4] was this: Nature is marked by ruthless, incessant competition for survival; to describe this idea, Darwin used phrases (that were to resonate long after) like "the universal struggle for life," "the struggle for existence," "battle within battle," "the great battle of life," "the war of nature," and "the great and complex battle of life."[5] The organisms locked in this struggle are not merely competitors, but enemies.[6] Crudely put, it is a zero-sum game gone haywire: Either you outlive your competition or you perish.

In January 1880, a few months before his death and twenty-one years after the publication of *Origin of Species*, the distinguished Russian ichthyologist Karl F. Kessler (1815–1881), rector of St. Petersburg University, chair of its Department of Zoology, and the first president of the St. Petersburg Society of Naturalists, delivered an address before a congress of Russian naturalists. His reverence for Darwin notwithstanding, he moved many a Russian biologist – but had little or no impact on the naturalists of Western Europe – with his concluding claim that "I obviously do not deny the struggle for existence, but I maintain that the progressive development of the animal kingdom, and especially of

[4] Full title: *An Essay on the Principle of Population as it affects the Future Improvement of Society, with Remarks on the Speculations of M. Godwin, M. Condorcet, and other Writers.* Malthus's core claim was that the growth rate of human population is geometric, whereas the growth rate of the food supply that is needed to sustain that population is only arithmetic. The latter would consequently seriously curb the former. Of his own doctrine of the struggle for existence, Darwin wrote, "It is the doctrine of Malthus applied with manifold force to the whole animal and vegetable kingdoms" (*Origin of Species*, 63). For Malthus's influence on Wallace, see Shermer, *In Darwin's Shadow*, especially 112–15.

[5] Darwin, *Origin of Species*, 62, 68, 73, 76, 79, and 80. Chapter 3, "The Struggle for Existence," in particular, details Darwin's views on this subject using a large array of examples.

[6] For example, Darwin, *Origin of Species*, 67, 69, 78, and 85.

mankind, is favoured much more by mutual support than by mutual struggle."[7] Among those convinced was Petr Alekseevich Kropotkin.

Kropotkin, in company with his naturalist friend I. S. Poliakov, explored the animal world in Siberia, especially the Vitim, Amur, and Usuri regions. Having read the *Origin of Species*,[8] they bore the book's claims vividly in mind as they journeyed off to do their fieldwork to test Darwin's theory. But, Kropotkin reports, they were unable to find that keen competition among animals of the same species that Darwin's magnificent work had led them to expect.[9] Instead, they found, as Kessler had taught, that while not everything in Nature was in harmony, it was remarkable how much of it was. There was an extraordinary number of examples of intraspecies as well as interspecies harmony, coordination, altruism, cooperation, and mutual aid and support.[10] And so, in his book *Mutual Aid*, Kropotkin wrote thus:

> But it may be remarked at once that Huxley's view of nature had as little claim to be taken as a scientific deduction as the opposite view of Rousseau, who saw in nature but love, peace, and harmony destroyed by the accession of man. In fact, the first walk in the forest, the first observation upon any animal society... cannot but set the naturalist thinking about the part taken by social life in the life of animals.... Rousseau had committed the error of excluding the beak-and-claw fight from his thoughts; and Huxley committed the opposite error; but neither Rousseau's optimism nor Huxley's pessimism can be accepted as an impartial interpretation of nature.[11]

[7] Quoted by Petr Kropotkin, *Mutual Aid*, 8.

[8] The Russian translation appeared in 1864 and was quickly sold out.

[9] Indeed, Darwin had claimed that competition between species of the same genera was far more intense than that between species of different genera; Darwin, *Origin of Species*, 76.

[10] The whole-heartedly anti-Malthusian Russian response to Darwin's theory of evolution and natural selection is neatly delineated by Daniel Philip Todes. For a synoptic view, see his paper "Darwin's Malthusian Metaphor and Russian Evolutionary Thought, 1859–1917"; and for a fuller account, see his book *Darwin without Malthus: The Struggle for Existence in Russian Evolutionary Thought.*

[11] Kropotkin, *Mutual Aid*, 9. Richard Dawkins would have fiercely denied this, dismissing it as "bad poetic science." Dawkins's position is that animals are neither essentially altruistic nor selfish. It is the genes that are selfish, and they, in the company of other genes, harness whole organisms in their service. Thus, whether organisms are selfish or not depends on whether this would have a salutary effect on the survival of genes; see Dawkins, *Unweaving the Rainbow*, especially Chapter 9, "The Selfish Cooperator," 212–14, 224.

Let this serve as a prolegomena to the problem of group rationality – namely, under what conditions is a group of scientists rational? – with which we are occupied in this book. Then, first, one might insist that the problem of group rationality is a problem whose solution must be sought in evolutionary terms, either in purely Hobbesian or purely Rousseauean terms – even if Kropotkin was right in admonishing that the truth resides somewhere in the middle. To offer a Hobbesian solution is to start with the premise that each scientist is interested only in his own domain of science; that is, each scientist identifies his own welfare with the welfare of his scientific domain. One then proceeds to show how these purely self-interested scientists could collectively reason themselves into a group of cooperating scientists. This I take to be the approach a game theorist might adopt (although, as we shall see, not only game theorists do); it is the approach I eschew.

Second, the Rousseauean approach may be far more interesting than the Hobbesian one. Here's why. The puzzling, momentously significant thing in biology was the discovery not only of the fact of evolution and the principle of natural selection but also of the *fact* of cooperation, altruism, and mutual aid and support among animals. It was a fact that for a long time remained unexplained in Darwinian terms. I should like to argue that the problem of group rationality is not to find out, at this late stage, whether there is, or ought to be, cooperation among scientists or not, or even whether the group is better off cooperating or not. *That* is surely a given. It is rather to discover what shall be the modus operandi of that cooperation. If he was minded, Kropotkin would have argued, as Paul Feyerabend was to do nearly a century later, that scientists should let the democratic method govern not only their fundamental political structures but also their scientific practices. This, then, is what the Rousseauean approach will make starkly clear: Given that they ought to cooperate, *how* should scientists do so? – *that* is the cardinal problem of group rationality.

Third, it should enable us to focus more sharply on just *what* the problem of group rationality devolves around. Nearly all philosophers of science – in fact, I believe, all of them – take scientific theory to be the prime element in dealing with this problem. They then try to understand how the scientific group should be structured around a theory or theories. My approach is markedly different. I shall argue that in order to determine the solution to the problem of group rationality, we ought to make method or methodology the cornerstone of our inquiry. This approach has other benefits, but it also has unintended, and sometimes surprising, consequences for the theories the group must pursue. For example, beginning with the methods that should structure a scientific society, we may be able to show, as an unintended consequence, that the society of scientists should proliferate theories. Consequently, this view of group rationality accommodates the earlier views that begin with theories rather than methods, yet the converse is not true; and hence, because of its depth, sweep, and generality – not to speak of its sheer elegance and beauty – this view of group rationality should have a much broader appeal.[12]

This chapter addresses three tasks. In Section I, it outlines the basic shape and substance of the main argument of the book. It does so by sketching the five reconstructed solutions to the problem of group rationality, revealing how the inadequacy of one solution leads to the next and how, in some instances, the virtues of one solution are preserved in subsequent solutions (with an occasional backward step). There is cross-fertilization, too: Later views are scanned in the light of earlier ones, revealing smudges. Another way of reading this book is to construe it as the unveiling of a budget of problems, separately marked and distinguished in this section, that a satisfactory theory of group rationality must solve. Section II then considers the question whether group rationality should be accorded priority over individual

[12] There is an important additional element that may introduce a wrinkle. "The virtue of Accuracy plays an important part in guiding and sustaining a collective division of epistemic labor.... and there is of course a genuinely historical story, a hugely complex one, of the cultural and eventually industrial sophistication of this idea into what is now called 'science.' One important feature of that process has been the way in which the understanding of nature itself affects what counts as an appropriate and effective division of labor." Thus Bernard Williams in *Truth and Truthfulness: An Essay in Genealogy*, 141.

If Williams is right, then our very understanding of nature will affect what will count as an appropriate and effective division of labor; methodology seems to play *no* role in this. The science of Anaximander will proclaim one division of labor in cosmological science, the science of Hawking a different division. Yet to ask whether either division of labor is effective, or which one is more effective, is not to raise a query in cosmology; it is a normative question pertaining to group rationality. Let us, then, accommodate Williams's point in this way. Science will tell us what disciplines and subdisciplines it will need. A theory of group rationality will tell us how the group is to be structured, or how the division of scientific labor is to be made, if the group is to function effectively and the disciplines and subdisciplines are to exhibit growth of knowledge.

rationality, or the other way around – neither notion being dispensable. Finally, section III details what I call the Williams problem. This is a problem of explaining the relationship between social structures and a solution to the problem of group rationality. Specifically, it asks what kind of ideal society (Social Utopia) would be needed to nurture the ideal scientific society (Scientific Utopia) envisioned in the solution. Conversely, the consequences, unintended or otherwise, of having a certain kind of Scientific Utopia for the shaping and form of a Social Utopia would make for an additional – hardly innocuous – way of evaluating a solution to the problem of group rationality.

I. The Plan of the Book

Let me sketch the spine and structure of the book. Simply put, the problem of group rationality is this: Under what conditions is a group of scientists rational? It is astonishing what a marvelous variety of problems can lie behind that seemingly simple formulation of the problem. It is with these various formulations of the problem that we shall be concerned in Chapters 2 and 3. In significant part, Chapter 2 will try to demonstrate that the problem of group rationality is a unique problem, neither reducible to nor analogous to a problem in game theory, social choice, social justice, or another such approach. After the uniqueness claim has been established, the problem of group rationality will be defined at the end of Chapter 3, and it is this formulation of the problem that will be deployed in the rest of the book. In attempting to gauge the adequacy of various solutions to that problem, we shall try to retain the integrity of each solution by first formulating the problem from the perspective of each of those solutions, in order to see how that particular formulation fares against our own. Once or twice, we shall also examine the adequacy of a solution in light of a different formulation of the problem in order to divulge the complexity and richness of the task.

Epistemologists, following Descartes, typically begin by delineating the skeptical position. Our starting point shall be no different. One begins with that position – and there is no other position antecedent to skepticism more challenging – with the aim of showing that such a position is either answerable, untenable, or contradictory (or even unworthy of a reply). Thus, the first solution we shall consider is that of the skeptic (Chapter 4); we shall see him offering both a negative and a positive solution. The negative solution will consist in arguing that a scientific group is rational provided the group is structured along the lines of *any* method that seems viable to *any* practicing scientist: no exceptions. This is famously captured in the slogan "Anything goes." The aim is to produce what the skeptic desires, namely, a vast, conflicting set of scientific theories, metaphysical outlooks, and methods of doing science. This view of the skeptic will be illumined by a tale. Intriguing as the skeptic's notion of a Democratic Council (wherein a lay person rules) may be, it is essentially an offshoot of his skepticism; thus, the skeptic's claim, reminiscent of the Greek sophists, is that whatever this council decides is epistemically right.

The positive solution consists in the skeptic's own favored method; of course, he makes no special plea on its behalf, claiming only that it is his preference. Let any scientist join in who has a similar preference; arguably, an enlightened laity will follow the skeptic's plan. Not only shall we find both these solutions – negative and positive – untenable, we shall also try to show how the skeptic's view is infected by a contradiction at its center (as if that should matter to the skeptic).[13]

Not less significantly, the skeptical position will present us with a range of problems that a theory of group rationality will need to solve. For example, should there be a single aim that informs a group of scientists? The skeptic argues that there should be a multiplicity of aims (without any, other than self-imposed, restrictions). Then, assuming that there is a multiplicity of aims, how shall we avoid the problem of fragmentation, namely, the problem of the group being splintered, each scientist going his own way, resulting in lost scientific labor? The skeptic would quarrel with our contention that scientific labor is lost, especially if scientists are engaged in doing exactly what they want and are "flourishing." Third, should there be a well-defined structure for the group? From the vantage point of the skeptic, clearly not.

[13] "Suppose that we show that some X he holds or accepts or does commits him to behaving morally. He now must give up at least one of the following: (a) behaving immorally, (b) maintaining X, (c) being consistent about this matter in this respect. The immoral man tells us, 'To tell you the truth, if I had to make the choice, I would give up being consistent.'" Robert Nozick, *Philosophical Explanations*, 408. What Nozick has a moral skeptic saying, we can have a skeptic in methodology saying (or something similar). See also, Chapter 3 of this volume, footnote 1.

These, then, are the cardinal problems that the skeptic's view leaves us to struggle with:

- What should be the scientific aim or aims of the group?
- How should the problem of fragmentation be dealt with?
- What should be the basic structure of a scientific group?

And so we are led to the second solution, the first version of the subjectivist view (Chapter 5). This solution shares one feature in common with the third solution, the second version of the subjectivist view, and with the fourth solution, the objectivist view (but, arguably, not with the fifth solution): namely, that a scientific group is rational provided the group is structured along the lines of a *single* method (that group structure should be defined in terms of method is explicit in none of the views). All these solutions, therefore, attempt to veer away from the skeptical view, and to that degree represent a small advance.

Now, a powerful dogma in methodology is the principle of proliferation. This principle – used as a yardstick against which to measure a theory of group rationality – states that there should be a proliferation of theories in a scientific society; a society nurturing a single theory, or an extremely limited number of theories, must provide a sharp defense for its practice. This first version of the subjectivist view claims that a society (the Rousseauean society of scientists) in which scientists aim to pursue truth and verisimilitude is unlikely to satisfy the principle of proliferation; but scientists interested in pomp, power, and circumstance (the Hobbesian society of scientists) will, inadvertently, satisfy it. This version of the subjectivist's view emphasizes subjective *non*epistemic values of the scientists that, supposedly, will take the group to where it objectively should be, epistemically speaking. Thus, this solution must, at the very least, explain (let alone justify) the tie between the subjective nonepistemic values of the scientists and the objectively viable theories that the group produces as a consequence of holding those values.

Consequently, some of the new problems this version of the subjectivist's view presents are these:

- What are the subjective, nonepistemic values of scientists?
- How should these nonepistemic values be distributed in the group with a view to satisfying the principle of proliferation?

- Can a single-method model sustain the required distribution of nonepistemic values as well as the plurality of theories?

The third solution is the second version of the subjectivist view. While the first version emphasizes the traditional notions of truth and verisimilitude, the second version will have none of that. Instead, it talks in terms of maximizing efficiency in puzzle solving. Like the first version, it too highlights the social, political, and economic structure of the society in which scientists do their science; having transformed our image of science, it would appeal to the history of science as a judicious arbitrator of competing theories of group rationality. Yet the relation of the history of science to these competing theories is a bit ambivalent. This solution to the problem of group rationality emphasizes epistemic values, claims them to be defining of science; but it leaves it up to the scientists in the group to decide what weights should be assigned to these values. Without much argument, it assumes that somehow this way of assigning weights and distributing the epistemic risks will lead not only to the acceptance of a single theory or paradigm (its preferred way of structuring the group), but also to the growth of knowledge (understood as greater puzzle-solving efficiency).

In this version, we shall also introduce a new, significant distinction between the static problem of group rationality and the dynamic problem. The static problem of group rationality is to determine whether a group of scientists, at a given time, is structured rationally; the dynamic problem of group rationality is to determine whether a group of scientists has evolved over time to a rational structure or to a more rational structure. Even if no solution discussed in this book draws that distinction – and whether one assumes that methods or scientific theories lie at the core of defining the structure of a society of scientists – the distinction is mentioned here in order to register its utter importance.

Finally, the second version of the subjectivist view employs the notion of "negotiation." To explicate this notion, a device akin to the Rawlsian notion of the original position is utilized. Scientists are placed therein, where they can, without prejudice or hindrance, negotiate with one another over what the structure of their society of scientists should be. This not only highlights the problem of group rationality, but also brings to the forefront, in a novel way, the problem of how to characterize individual rationality and its connection with the problem of group rationality. Thus, this version leads to some fresh, significant problems of its own making:

- What role should be accorded to the social, political, and economic structure of the society, as well as to the history of science, in developing a normative theory of group rationality?
- What makes a group of scientists statically rational, or rational at a given time?
- What makes a group of scientists dynamically rational, or rational over an interval of time?
- How might the notion of negotiation be conceived and connected to the reasons that individuals scientists offer in arriving at a plausible theory of group rationality?

The fourth solution is the objectivist view (Chapter 7). It is a marked improvement over the foregoing views in that, unlike the skeptic's view, it reckons the possibility of a genuine solution: a solution that, unlike on the subjectivist's view (both versions), can function as a criterion against which to measure subjective epistemic values themselves, and thus the rationality of the group (keeping this assumption sharply separate from a stronger one, namely, that the objectivist view *has* a solution that *is* objectively right; in my view, the solution isn't right). The objectivist view also prominently claims that a scientific group is rational provided the group is structured along the lines of a single best method. While, as we shall see, this is seriously inadequate – for one thing, it is too optimistic – it brings to the fore a powerful problem in the field of meta-methodology. This problem is best explained by analogy to the problem of demarcation.

The problem of demarcation is, "What distinguishes a scientific theory from a pseudo-scientific one?" The problem was deemed quite significant in philosophy of science, for reasons that are old and well known. Now, if the subjectivist views as well as the objectivist view of group rationality insist that a single method be used to structure the society of scientists, and given that there are a fair number of methods available, there are two nice problems to be dealt with – and not just for these views. The first is the new problem of demarcation: What distinguishes viable methods from those that are not? The second problem is: Of the viable methods, which is the best method? Since we want the group of scientists to be structured or grounded along the lines of the best available method, the solution to the two problems is a sine qua non for any solution to the problem of group rationality. This, I trust, shows, in a rather intriguing way, how meta-methodology (the evaluation of methods) and methodology (a theory of group rationality, among other things) are inextricably bound.

The objectivist view, then, harvests the following problems:

- How should a viable method be demarcated from a nonviable one?
- How rational is a group of scientists structured along the lines of a single method?
- Given the deep underlying interlacing between method and meta-method, how shall we account for the growth of method (and not just for the growth of science)?

The penultimate chapter, Chapter 8, will examine the work of Hilary Putnam. Putnam's recent views attempt to demonstrate a remarkable parallel between the scientific image and the moral image, and in so doing Putnam delves deeply into what makes for individual rationality. Many moral images – constrained by the principle of equality and freedom – should adorn a society, says Putnam; might Putnam also claim that as many scientific images bedeck a society of scientists? What would serve as constraints on these images? The chapter will examine, too, the consequences that Putnam's view of individual rationality will have on group rationality; and it is here that Peirce's puzzle will play a central role. Suppose that an individual has to choose between two courses of action. If the probability of success of doing one action is higher than the probability of success of doing the other, why – Peirce's puzzle – should, the person do that action that has the higher probability of success? Any solution to the problem of rationality, which derives from the assumption that the ultimate ground for being reasonable is that one will arrive at truth in theory, and success in action, more frequently if one is reasonable, is untenable. So Putnam claims, and then offers a solution. The outcome of that inquiry will instruct us to hone and shape our theories of individual rationality differently, if we are not to succumb to relativism (a threatening prospect). One might adopt the Kantian view as an alternative; and this, I aver in a programmatic note, might enable us to reap benefits denied us on theories examined in this book.

Here, then, are three crucial problems that Putnam's view will bring to the fore:

- What is a scientific image, and what scientific images may govern a society of scientists?
- What is individual rationality? Or, how shall Peirce's puzzle be solved?
- What is the role of individual rationality in a theory of group rationality?

Each view of group rationality half-hides – sometimes not that successfully – a view of Scientific Utopia that it would promote, and each portrays, by implication, a scientific dystopia.[14] Consequently, we shall examine the nature of that utopia that is envisioned by each theory of group rationality and ask, in a manner after Bernard Williams, "What kind of social and political structure will be needed to create and sustain that Scientific Utopia?" Williams was hyperalert to the possibility that a seemingly plausible moral theory (such as utilitarianism) could be implemented only in a rather gruesome or high-handed way. He took that to show the defect of a moral theory. Williams offers us, then, an additional way in which we can criticize a normative theory of group rationality: If a normative theory of group rationality implies a social and political structure that is unacceptable from a moral point of view – the Social Utopia it requires to flourish cannot possibly strike our fancy – then it is not a normative theory that can appeal to us. (Some may see it as an occasion to revise our vision of a Social Utopia.) By implication, one significant task is to show how a Scientific Utopia implied by a theory of group rationality can coexist easily with a Social Utopia; they might even be shown to be mutually reinforcing.[15]

[14] To keep to manageable limits, I say little about dystopia; but I trust that what is said about utopia will leave little to the imagination about what forms of dystopia will result, and why they will be deemed unacceptable.

[15] It might be feared that I am attempting to construct, perhaps even to legitimize, conditioning Scientific Utopia on Social Utopia. I hope it will be abundantly clear in what follows that I give primacy to the solution to the problem of group rationality. Once that solution is at hand, then and only then may we ask what kind of Social Utopia will be needed to sustain that rational society of scientists. Like Williams, however, I do not rule out a priori the possibility that the Social Utopia that will be needed would be so bizarre or frightening that we might wonder whether we have found the right solution – and, if we have, whether we are morally obliged to bring about a Social Utopia in which that solution can be implemented. I cannot imagine any such obligation.

Perhaps this is just another version of Putnam's idea of trying to produce a coherent picture out of scientific and moral images.

Finally, we might say that Williams has presented us with these cardinal problems:

- What Scientific Utopia is implied by a given theory of group rationality?
- Under what kind of Social Utopia could that Scientific Utopia sustain itself and thrive?
- Would that Social Utopia be morally acceptable?

Shaped and colored by the arguments of the book, the final chapter will present various problems – nine, to be exact – related to the cardinal problem of group rationality, and the intricacy of their connections to each other, from a distinctive point of view, the better to enable us to offer a richer, deeper, and a more unified theory of group rationality.

II. Group to Individual, or Vice Versa?

The problem of group rationality – "Under what conditions is a group of scientists rational?" – might be offered under two different guises. We shall call them the *Group-to-Individual Problem* and the *Individual-to-Group Problem*, respectively. The Group-to-Individual Solution attempts to show how individual scientists collectively decide how their group should be organized and structured and how, having thus decided, each member acts in accordance with that covenant. The scientists decide what should be the ultimate goal (say, truth or verisimilitude) toward which their group should be aiming (and one of the original problems may well be how to settle on that ultimate group goal). Next, given the goal, they determine what should be the structure – defined by method(s), I shall assume (contrary to the prevailing assumption) – of their group, and how that method or those methods should be selected. Finally, each scientist decides how best to act in order to comply with the foregoing two determinations or decisions. In this Group-to-Individual Solution, the group is accorded priority over the individual scientists who compose the group. Indeed, the identity of the group or subgroup, defined in terms of method(s), may well remain tolerably constant while the individual scientists move in and out of subgroups, or move out of the group altogether. The knowledge of the scientists about the group's goal and method(s) determines their subsequent actions. It is no mystery if their collective actions produce a harmonious result; it was intended. It is this approach that infuses this book.

On the other hand, the Individual-to-Group Solution is more ambitious and, perhaps, more elegant. It takes no overall picture of what the group is about or what the group aims at as its starting point. Rather, each scientist in the group acts in accordance with what his or her interests, aims, and goals dictate, thus primarily relying on individual rationality – seemingly a crisper, clearer notion than group rationality. The approach shows how these independent (although often hardly independent at all) actions of individual scientists collectively lead the group to be ideally or rationally organized, structured. It is a view that one might read into Hilary Putnam.[16] This would be another marvelous illustration of spontaneous order, courtesy of unintended consequences. In this Individual-to-Group Solution, committed to methodological individualism, the individual is accorded priority over the group. The identity of the group, or even its structure, is unimportant; the knowledge that scientists possess about their own ultimate interests is sufficient not only to determine their actions, but also to produce a spontaneous order, one that was consciously intended by no one.[17]

Two remarks about the Individual-to-Group Solution: First, the Individual-to-Group Solution invokes only norms, norms of individual rationality, and on the basis of those norms shows how it is rational for each individual to act in the way he does, thereby conjointly producing a rational group. Nevertheless, it seems quite unlikely that individual scientists acting independently will produce a spontaneous order exactly in conformity with the norm for the group. Too many things have to come together, without plan or design, to produce that result, reducing its probability markedly. But second, even if such a solution were had – it would be fascinating to have it – the Individual-to-Group Solution would still need the Group-to-Individual Solution because it fails to give an independent account

[16] See Chapter 8 of this volume.
[17] For example, see Cristina Bicchieri, *Rationality and Coordination.*

of how a group *ought* to be ordered; or, at the very least, it must presuppose that solution.[18] One can regard the Individual-to-Group Solution as a way of *explaining* how an ideal group can come about (an invisible-hand explanation, so beloved of Adam Smith and Robert Nozick), but not as *justifying* the normative correctness of the solution. That task must be left to the Group-to-Individual Solution. In either case, I conclude, the Group-to-Individual Solution is either favored or indispensable.[19]

[18] For a third argument, from a vastly different perspective, against the possibility of the Individual-to-Group solution, see Chapter 9, section I, this volume.

[19] Let us say that where each scientist acts rationally to maximize his or her welfare, there is perfect coordination of plans of scientists, and preeminence is given to those patterns of collective action that constitute an equilibrium. But several equilibria cannot be based on individual rationality alone. What is further required, says Cristina Bicchieri, is prediction about what other scientists will do, which will influence the actions of the predictor. Then, to be able to predict correctly what other scientists will do, it is necessary to assume that the predictor's beliefs about what other scientists will choose are both consistent and true. Consequently, what is needed is a true description of how beliefs are formed and how the changes in beliefs that lead scientists to have true beliefs come about.

Assume that such rules of belief formation and revision are available and that these rules specify a criterion of equilibrium selection; if this criterion identifies a unique equilibrium as the solution for the game, then scientists who have common knowledge of (*a*) rationality, (*b*) shared rules for belief revision, and (*c*) a shared criterion of informational value can identify their equilibrium strategies. It is "important to have a theory of the game that includes a description of players' knowledge of the theory itself. A good theory of the game will be self-fulfilling theory." Bicchieri, *Rationality and Coordination*, x.

One problem with such ways of viewing group rationality is this: If a good theory of the game already provides each scientific player with a norm about group rationality (which of the several equilibria is the best), and regardless of whether too much knowledge produces inconsistencies, then it is not clear how merely individual rationality (plus the shared rules for belief revision, the shared criterion of informational value, and so on) would *alone* suffice for the task of determining group rationality. Granted: Once such a normative theory of what the group norm should be is presupposed, *then* it is possible to show how each scientist, having accepted such a group norm, would act in view of it, and why it is to his advantage to do so; and the rest would follow.

Nor would the following serve as a defense: Over an interval of time, the scientists would evaluate (employing only individual rationality, etc.) the equilibria that are available, and settle on the ideal one. This will not do, because it already presupposes the equilibrium they should aim at, and it is in terms of that equilibrium that scientists are made to negotiate and evolve a solution. My purpose here has been no more than to indicate that hidden behind such invisible-hand explanations lies the norm of group rationality, in the absence of which that explanation would fail. Having such a norm, then, is at least a necessary condition.

Since logical possibilities are stubborn pests and cannot be eradicated, let me conclude on a cautionary, compromising note. First, it is entirely possible that once a normative theory of group rationality is provided, one might be able to offer an evolutionary explanation of why just those norms that ought to govern the group do, or can, prevail in the group. (An organism ought to develop an eye, and so it did.) But if we are to be mindful of Hume's Fork, then the evolutionary explanation may provide uncanny support for the former theory by showing that although what is recommended as a norm is a realistic possibility, it cannot be offered as a substitute for the normative theory. One is about what *is* the case, the other is about what *ought* to be the case. Thus, even if such an evolutionary theory is ultimately available, it cannot dispense with a normative theory of group rationality. At best, there will be a symbiotic relationship between the two.[20]

Second, one might be able to offer a game theorist's account of how such a rational group of scientists could come about, making plausible assumptions about the rationality of an individual scientist or actor. This would be a powerful result – perhaps someone at RAND Corporation may produce it. Even so, two points are worth noting: First, unless one produces a normative theory, there is nothing for the game theorist *to* demonstrate, either that the normative theory is true or that it is false. By way of analogy: Some game theorists argue that there are serious problems with the Rawlsian notion of the original position; and if so, then John Rawls's (1921–2002) theory of justice might be suspect. Rawls had argued that in the original position a moral agent does not possess knowledge of a variety of significant things about himself, things that are arbitrary from a moral point of

[20] Here it is useful to recall what Thomas H. Huxley said (and what, in our day, Stephen Jay Gould used to admonish): "And, again, it is in error to imagine that evolution signifies a constant tendency to increased perfection. That process undoubtedly involves a constant remodelling of the organism in adaptation to new conditions; but it depends on the nature of those conditions whether the direction of the modifications effected shall be upward or downward. Retrogressive is as practicable as progressive metamorphosis." *Evolution and Ethics and Other Essays*, 199. Had Thomas Kuhn kept this remark in mind as he wrote the last few pages of *The Structure of Scientific Revolutions*, that book might have been rather different. (See Chapter 6 of this volume, especially sections I–III.) *Perfection* as an evaluative term, then, is no more an evolutionary notion than *group rationality*; a group's retrogression can just as well be explained in evolutionary terms as its progression. Clearly, we will need to keep evolutionary explanation disentangled from normative justification.

view, such as his natural talents, wealth, religion, and place in the social, political, and economic structure of his society. The game theorist's argument might go that if the agent knows nothing that is morally relevant, it should not really matter to him what he decides; he is in no position to judge the consequences that matter to him; hence, he is in no position to make any decision. Or he might choose one option just as well as any other. (One might ask, not entirely rhetorically, "Should one then provide an agent with information that *is* arbitrary from a moral point of view?" And, if so, why should such a decision then be deemed without prejudice?)

Give the moral agent some information, and one can construct a game theorist's approach to the issue of social justice. For example, if justice commands equality, so this argument goes, the game theorist can show how the parties to the dispute will reason themselves into accepting the principle of equality.[21] My contention is that even if this argument is sound – and that is conceded only for the sake of argument – it would be hard to overrate Rawls's achievement for the insight it has offered into the idea of the principle of equality. In the absence of Rawls's principle of equality, the game theorist would not have known *what* his game theory model is supposed to accomplish or explain – or even *what* would count as his game theory model having done better than Rawls's principle of equality. Similarly, then, a sufficiently powerful normative theory of group rationality, even if mistaken, may lead to an insight or two; in its absence, a game theorist would not know what his model is supposed to explain.

Third, even if the end results match, what is by far more significant is the reasoning that leads to those results. If Rawls's vision of a just society is right, the reasons that lead the moral agents to a covenant that marks a just society are quite different from the reasons that would lead a moral agent in, say, a Prisoner's Dilemma to the same covenant. Perhaps there might be structural isomorphism in the reasoning in the two cases, but that will have to be demonstrated (a substantial task, intrinsically interesting in its own right), not assumed. In a parallel way, the end results of a normative theory of group rationality may well be produced in game theory, but the reasoning that leads to the results

[21] The issue of how precisely the principle of equality must be qualified is not germane here.

in the two cases may diverge remarkably, unveiling different aspects of rationality (both individual and group), none of them reducible to any other.

III. The Williams Problem and Utopias

Bernard Williams (1929–2003) was a powerful exponent of the idea that ethical considerations alone do not provide the necessary *force* for a moral agent to do what those considerations dictate. His pro-Humean stance insists that reason is impotent to get us to act ethically; *sympathy* is at least a necessary condition.[22] What is also required are appropriate psychological states or dispositions coupled with an appropriate set of social, economic, and political institutions that will enable those ethical considerations to be inculcated and perpetuated in society, engendering *confidence* – Williams's term of art. In the absence of such dispositions and institutions, ethical knowledge, and the possibility of acting on that knowledge, disappear.[23] Earlier, in his short classic, *Morality: An Introduction to Ethics*, Williams had argued that Plato, unlike Aristotle, foresaw clearly what would happen if the artists were given free rein. Plato feared that such freedom would inject instability into an otherwise well-ordered moral system – hence the banishment of the artists from society in order to ensure a stable moral order, an order imposed upon the rest by the philosophical guardians.[24]

Here is what Williams said in his later, skeptical book, *Ethics and the Limits of Philosophy*. Some have averred that

> the justification of the ethical life could be a *force*. If we are to take this seriously, then it is a real question, who is supposed to be listening. Why are they supposed to be listening?[25]

And three pages later, this:

> Plato who saw more deeply than any other philosopher into the questions raised by the possibility of a life outside the ethical, did not himself take it for granted that a justification of the ethical life would be a force. He thought that

[22] Bernard Williams, *Morality: An Introduction to Ethics*, 12.
[23] Bernard Williams, *Ethics and the Limits of Philosophy*, especially 147–8 and 170–1.
[24] Williams, *Morality: An Introduction to Ethics*, especially 57.
[25] Williams, *Ethics and the Limits of Philosophy*, 23.

the power of the ethical was the power of reason, and that it had to be *made* into a force. He saw it as a problem of politics, and so it is.[26]

Arguing by analogy: Offering a theory of group rationality, a normative theory, would not in itself induce scientists to band together to form a rational group. The normative theory of group rationality, one might say, has no force. It is merely the power of reason; it has to be *made* into a force and thereby made into a problem of politics.[27] *How* can it be made into a force? Appropriate psychological states or dispositions coupled with appropriate social, economic, and political institutions (Social Utopia) must prevail in which a society of scientists structured in a particular way can flourish (Scientific Utopia); and scientists who negotiate and accept that structure must make decisions to accept or reject scientific theories in light of the method that defines that structure. Indeed, even the moral worth of science would have to be negotiated.[28] *Only then* will that structure be implemented; only then will the power of reason become a force.

Let me, then, label the following set of questions as *the Williams problem*:

What is the ideal social order (Social Utopia) presupposed by a theory of group rationality? Is that Social Utopia pliable and morally justified? What is the ideal scientific society structuring the practice of scientists (Scientific Utopia), presupposed by that same theory of group rationality? Is that Scientific Utopia realistic and how successful would the scientific practice be in that utopia? Can the Social Utopia lie in cohesion with the Scientific Utopia?

[26] Ibid., 26–7; see also *Morality: An Introduction to Ethics*, especially 57–8, and Williams, *Truth and Truthfulness*, 69.

[27] "In my sense, to be skeptical about ethics is to be skeptical about the force of ethical considerations; someone may grant them force, and so not be a skeptic, but still not think that they constitute knowledge because he does not think that the point lies in their being knowledge." Williams, *Ethics and the Limits of Philosophy*, 25. I do not share Williams's pro-Humean stance in the moral field. Williams's pro-Kantian stance in philosophy of science is a different matter; see Williams, *Ethics and the Limits of Philosophy*, Chapter 4. Now, what I say here should appeal to both Humeans and Kantians. There are two distinct problems on either view. First, what is the right theory of group rationality? and second, given the answer to that question, how shall we implement that theory in society? Williams's view deserves close attention because he sharply highlights the social and political implications of a normative theory.

[28] Williams, *Morality: An Introduction to Ethics*, especially 58–9.

Williams demands that "We should ask a pretended justification [of the ethical life] three questions: To whom is it addressed? From where? Against what? Against what, first of all, since we must ask what is being proposed as an alternative."[29] (In the moral case, a justification for being ethical is arguing against the motivations of an amoralist.)[30] By parity of reasoning, if we are attempting to justify a theory of group rationality, then Williams would have us ask three similar questions. Our answers to these questions would have to be: First, it is addressed to scientists generally, and not to scientists in any particular discipline or subdiscipline. (This would answer the question "To whom is it addressed?" or "Who is supposed to be listening?") This leaves entirely open answers to two fairly significant questions: (*a*) What are the appropriate psychological states or dispositions of the scientists; and (*b*) how are these states or dispositions to be connected to the reasons that ought to persuade them of the correctness of a theory of group rationality?

Second, the justification of a theory of group rationality would come from reason and practice; eventually, the success of the scientific enterprise must be revealed in the practice of the scientists. "Practice is primary," says Hilary Putnam. (This would answer the question "Why are they supposed to be listening?" They are supposed to be listening because their scientific success depends upon it.) We would have to have a measure – a methodology – by which to gauge the success of the practice of science, that is, to gauge whether science (domains or subdomains, singly or collectively) is progressing or declining. This gives another powerful reason why the structure of a society of scientists should be defined in terms of methodology.

Third, and hardly least important, a theory of group rationality is not attempting a solution from nothing – it is not a view from nowhere – but rather is pitched against other theories of group rationality. These other theories provide the backdrop against which a rival theory can be seen as arguing. (This would answer the question "Against what?") In this book, we shall see the skeptic (on one reading) demanding a

[29] Williams, *Ethics and the Limits of Philosophy*, 23. Once again, Williams is raising the issue in the context of an ethical theory; I am using his potent questions to explore a different normative field: the field of group rationality.

[30] Williams, *Ethics and the Limits of Philosophy*, 26, and Williams, *Morality: An Introduction to Ethics*, 3–13.

justification for needing a theory of group rationality in the first place, putting on notice all purported theories of group rationality. Or the skeptic might be seen (on another reading) to offer "Anything goes" as the ultimate dogma that should govern any society of scientists, again undermining any alternative "law-and-order" theory. Next we shall follow the subjectivist view of group rationality (both versions), a view that can also be read as an answer to the skeptic; then the objectivist view will be presented as an alternative to the subjectivist view; and finally we outline Putnam's view of group rationality – not without having shown the significance of individual rationality – against some extant theories; it will be especially interesting to see if Putnam has answered the skeptic. And thus we shall have come full circle.

One final analogy: John Rawls had argued, in *A Theory of Justice*, that formal constraints must be placed on principles of justice; any principle that violates any of these constraints must be unacceptable. One of these formal constraints is the publicity condition. "The parties assume that they are choosing principles for a public conception of justice. They suppose that everyone will know about these principles all that he would know if their acceptance were the result of an agreement. Thus the general awareness of their universal acceptance should have desirable effects and support the stability of social cooperation."[31] Williams recoils from the publicity condition – he calls it the *transparency* condition – on the ground that it is one thing to maintain that social and political relations should not be entered into through ignorance or misunderstanding,[32] quite another to insist on entering those relations only on explicitly stated principles of justice.[33]

The publicity condition may well have a significant role to play in a theory of group rationality, especially one that casts methodology in a primary role. Michael Polanyi, siding with Williams, might argue that just as ignorance and misunderstanding should not be the basis of social relations, so, correspondingly, permitting the incorrect

[31] John Rawls, *A Theory of Justice*, 130–1; see also footnotes 5 and 8.

[32] One might say that Williams's relentless criticism of utilitarianism – at least Government House utilitarianism – is intended in part to show that it violates just this part of the condition. See *Ethics and the Limits of Philosophy*, especially 109–10; see also "The Point of View of the Universe: Sidgwick and the Ambitions of Ethics," in *Making Sense of Humanity*, especially 165–7.

[33] *Ethics and the Limits of Philosophy*, 101–2.

reporting of experimental results, or providing fudged data, and so on should not be the basis on which to form a cohesive scientific group. Yet, far more significantly, in order to form a cohesive group it is not essential to explicitly state the principles of a method or methods around which the group is structured. Perhaps it is not even possible to do so, because there is a tacit dimension of this knowledge that guides the scientist; it is the feel for the discipline he acquires, nurtured through long practice, that the scientist cannot articulate and that escorts his activities and decisions – not some formal, objective rules formulated by a method – about which theories to pursue, to regard as holding promise, or to let fall by the wayside.[34]

By contrast, there is the Rawlsian picture of scientists gathered in an original position of their own[35] trying to determine what would be the best and most effective way of structuring their group. Each proposal – namely, the methodological principles – for structuring the group is considered. Consequent upon an agreement between scientists over what method(s) should structure their group, the publicity condition will facilitate the general awareness of the universal acceptance of that method or those methods, and that awareness will also produce desirable effects and stability of scientific cooperation. It will have one additional virtue, *criticizability*.[36] That which is explicitly stated, and widely

[34] See Michael Polanyi, *Personal Knowledge: Towards a Post-Critical Philosophy* and *Tacit Dimension*. The following sentiment – concerning the intuitive condition of conflict of values that a person can live with privately in a way in which public order cannot or ought not – is made in the context of morality, but it is one which unerringly mirrors what Polanyi says in the context of science: "For the intuitive condition is not only a state which private understanding *can* live with, but a state which it must have as part of its life; if that life is going to have any density or conviction and succeed in being that worthwhile kind of life which human beings lack unless they feel more than they can say, and grasp more than they can explain." Bernard Williams, "Conflicts of Values," in *Moral Luck*, 82.

[35] This idea is more fully explored in Chapter 6, section IV, this volume.

[36] There is one fairly significant difference that for now I bury in a footnote. Rawls has claimed that whatever agreements are arrived at by the moral agents in the original position are agreements made in perpetuity. In other words, the principles of justice once agreed upon in the original position are not renegotiable or otherwise revisable; see Rawls, *A Theory of Justice*, 153. The view presented here is more along the following lines: An agreement between scientists is reached in the original position. They emerge from this position and conduct their scientific practice in light of their agreement. Over a significant interval of time, their practice reveals the drawbacks of their agreement or theory of group rationality. They reenter the original position, with that knowledge, and arrive at an agreement that is a better way of structuring their

acknowledged, can be publicly criticized and evaluated; such criticism and evaluation will make possible a more effective structuring of a group of scientists. In short, it would make possible the growth of reason. This bestows upon the Rawlsian view – at least as applied to the problem of group rationality – more plausibility than the alternative view of tacit knowledge of theory evaluation. Consequently, I analyze the various solutions to the problem of group rationality *as if* they relied upon, or presupposed, the publicity condition.[37]

This overview before us, let us begin our task. Let us begin with analyzing the problem of group rationality and ask if it is a unique problem and, if so, whether prima facie the problem itself can be formulated in a variety of ways. The next two chapters will answer those two questions in the affirmative.

group. They emerge again from the original position, practice their science, and the process of conjecture and refutation repeats itself, indefinitely. Consequently, a theory of group rationality is regarded as a conjecture that is criticizable. Undoubtedly, an enormous amount of work needs to be done to make this idea precise.

[37] Nor am I convinced that Williams would object. While Williams is diffident about the possibility of the growth of moral knowledge, he is not skeptical about the possibility of the growth of scientific knowledge. Indeed, he speaks about the absolute conception of the world and the possibility of knowing it; see Bernard Williams, *Descartes: The Project of Pure Inquiry*, 236–49 and 298–303, and Williams, *Ethics and the Limits of Philosophy*, especially 138–40. For an insightful commentary on Williams's notion of the absolute conception, see Hilary Putnam, *Renewing Philosophy*, Chapter 5. The interesting question here is: Given his conception of science, would Williams have found the use of the publicity or transparency condition congenial in forming and sustaining a society of scientists? I think so.

2

Group Rationality

A Unique Problem

"The two persons iterated Prisoner's Dilemma," says Robert M. Axelrod, "is the *E. coli* of the social sciences." An analogue of the Prisoner's Dilemma is found in a variety of fields, ranging from evolutionary biology and arms control policy to networked computer systems and political philosophy.[1] Consequently, when unrelated fields have analogous structures, the results in one are deemed applicable to the other. It might appear, then, that the problem of group rationality, namely, "How should scientists cooperate with one another to form or organize an effective group?" is so remarkably similar, say, to the problem of social justice, "How should individuals cooperate with one another to form or organize a just group?" or to the problem in evolutionary theory, "How do species reach an equilibrium state with neighboring species?" that one might unsuspectingly conclude: A little tinkering with an adequate theory of social justice or evolutionary theory would produce an adequate theory of group rationality.[2] Nothing is further from the truth. Or so I shall argue.

[1] Robert M. Axelrod, *The Complexity of Cooperation: Agent-Based Models of Competition and Collaboration*, xi.

[2] "Science is often seen as a zero-sum game. To the extent that science may be modeled as a game with rules, modern game theory may grant us a deeper understanding of the tension between competitiveness and cooperation by distinguishing between zero-sum and plus-sum models." Shermer, *In Darwin's Shadow: The Life and Science of Alfred Russel Wallace*, 142.

Here is what I shall attempt in this chapter. The problem of group rationality, I want to show, is a unique problem. It is not a problem that can be solved in evolutionary terms (section I), nor is it a problem solvable by the resources of game theory. Since this latter is a significant task, I address it in two sections (sections II and III), each section dealing with a distinctive treatment of the analogy between game theory and group rationality. That the problem of group rationality is not a problem analogous to the problem studied in economics, the problem of welfare economics, deserves to be treated separately, and so I address that task in the following chapter. My conclusion is a bit modest, for all that. Perhaps, eventually, someone might find a solution to the problem of group rationality in terms of, or analogous to, say, evolutionary theory, game theory, or welfare economics. Even so, that task can be accomplished only by *first* and *independently* showing what the solution is in substantive philosophical terms, and only then by casting it in these other terms.

I. Not an Evolutionary Problem

Let us begin with fireflies.[3] Wherever fireflies congregate, and they could be stretched over miles, they have been observed (in Thailand and Africa) to blink in unison; it is such an incredible phenomenon that in 1917, in *Science*, Philip Laurent wrote that "for such a thing to occur among insects is certainly contrary to all natural laws." Between 1915 and 1935, *Science* published twenty articles on the subject, but none could satisfactorily explain the phenomenon. To some it seemed as if there was a central command that issued directives that alone would explain this incredible synchronicity in blinking. Steven Strogatz and Rennie Mirollo conjectured that the rhythm was controlled by an internal, resilient oscillator in each firefly. Every firefly is sending as well as receiving signals from every other, thus enabling each firefly to mark the change in the flashing rhythm of other fireflies. What is more, the internal oscillator then adjusts its own flashing to coordinate with the flashing of other fireflies, thus producing this marvelous

[3] My account is drawn from that engaging book *SYNC: The Emerging Science of Spontaneous Order* by Steven Strogatz, Chapter 1, "Fireflies and the Inevitability of Sync."

spontaneous order.[4] What Strogatz and Mirollo proved in 1989 is that under specified conditions, this synchronicity cannot fail to occur.

What is philosophically significant about this example for the problem of group rationality is that, notwithstanding the beauty of the end result, there is no biological payoff in this joint venture. Not a biologist, I venture the thought that fireflies that are not part of the group that eventually produces this delightful result are not necessarily at an evolutionary disadvantage, nor are those that are part of this group thus evolutionarily better off. Hence, while the cooperating fireflies produce a happy result, it is not the *kind* of result we are looking for when we're seeking to show why cooperating is better than not. We want the result to be important in its own right, or to have some significance beyond its breathtaking order, symmetry, or beauty, produced by chance or design.

In short, the following would not help. Wherever there are scientists, they congregate in groups; that tendency to congregate is controlled by an internal, resilient oscillator or instinct in each scientist. Each scientist is both a sender and a receiver of signals to and from every other scientist, enabling each scientist to record the changes in activity of other scientists, thus allowing the internal oscillator of each scientist to adjust its own activity to the activities of other scientists, thus resulting in spontaneous group ordering of activities.[5] Let us even suppose that under specified conditions, this synchronicity *must* occur. While this would explain the phenomenon of coordination of scientific activities, it would leave unexplained, as in the firefly case, *why* the scientists coordinate their activities. But if one is seeking to answer the latter question, then postulating an internal, instinctive mechanism is utterly insufficient.

[4] Both terms, "spontaneous" and "order," are ambiguous. Fireflies, and cardiac cells fluctuating in harmony to produce heartbeats, may be far, far more spontaneous than other orders, such as the solar system. Beehives and ant colonies may be remarkably well ordered; cities are less so. Eighteenth- and nineteenth-century biologists were ordered in societies that by current standards might be deemed lax, relaxed, or loose. Following the herd, I shall make no attempt to make these notions precise.

[5] Notice how *much* is tacitly conceded: In the case of the fireflies, our focus was on one activity, namely, blinking. In the case of scientists, I have used this rather vague word *activity*. If one were to make that activity more specific – say, the activity of seeking an understanding of solar flares, market fluctuations, or Parkinson's disease – the explanation would falter much more quickly than it does now.

So let us turn to other examples – the animal kingdom is filled with them – in which the results of cooperation are, evolutionarily speaking, very significant. Let me delineate one famous example: vampire bats (*Desmodus rotundus*). In 1983, subsequent to the publication of the famous paper "The Evolution of Cooperation," by Robert Axelrod and William Hamilton, Gerald Wilkinson reported a fascinating discovery he had made in Costa Rica concerning vampire bats. Vampire bats inhabit tree hollows by day and scavenge for blood by night, making small, surreptitious incisions in skins of large animals and then extracting blood therefrom. This is risky venture, and a bat may return without blood-food for various reasons. Older bats are unsuccessful one night in ten, younger bats one night in three. A bat that goes without food for sixty hours – not an unusual prospect – gets close to death by starvation; death by starvation occurs in seventy-two hours.

To stave off this danger, cooperation is clearly to everyone's advantage. Now, bats often return to the hollows with more blood than they need and so can become potential donors (by regurgitating excess blood) to bats who have failed in their scavenging that night. There is a potential disadvantage to the fortunate bat in being a donor, because the recipient may fail to reciprocate; the potential advantage in being cooperative and altruistic now, even to a bat to which the donor isn't genetically related (although sometimes it is), is that the donor bat will receive help if it needs it in the future. In order for the cooperation between bats to work, there must be, and there is, a way of identifying the altruists and those who have reneged. This makes possible, as the phrase goes, tit for tat. To avoid such threats, vampire bats fall in line: cooperation rules.[6]

[6] Similar cooperative enterprises can be found among tree swallows (*Tachycineta bicolor*), male olive baboons (*Papio anubis*), African vervet monkeys (*Cercopithecus aethiops*), dwarf mongooses (*Helogale parvula*), stickelbacks (*Gasterosteus aculeatus*), the black helmet (*Hypoplectrus nigricans*), a hermaphroditic coral reef fish, and forty-five species of fish and at least six species of shrimp that offer to clean the big fish that come in from the open ocean (some even changing their colors to signal that they have come to be cleaned). So powerful are the cooperative instincts honed by natural selection that, in a case reported by Robert Trivers, a fish raised in an aquarium for six years, snapping up every fish thrown to it, opened its mouth to let in a cleaner fish even though it had no parasites it needed to get rid off. For technical details, see G. S. Wilkinson, "Reciprocal Food Sharing in the Vampire Bat" and "The Social Organization of the Common Vampire Bat." Recent research confirms even more: Despite a lack of bloodline connections, male vampire bats cooperate in a similar

Of course, there are always exceptions, or at least the appearance of exceptions.[7] Consider the puzzling example of the cuckoo.[8] A cuckoo invades a nest and gets the host and its offspring to sacrifice themselves for its benefit; it would appear that evolution has gone awry in this case. There is no advantage to the hosts, only acute costs; the cuckoo is a huge beneficiary, whatever limited costs it has to bear. Thus – or so it would seem – the hosts' cooperation is difficult to explain from the point of view of evolution.

Not so, says Helena Cronin:

> Indeed, it might seem that on this view we should positively expect the hosts to hit back. After all, there's nothing in it for them – in fact, it's a downright sacrifice, all give and no take. It seems little short of a Darwinian scandal for natural selection to allow the cuckoos their success. But our indignation would be misplaced. We shouldn't look on the hosts as systematic losers, even if they are condemned never to shake off their oppressors. There may well be an asymmetry in the strength of the selective forces acting on the cuckoos and their hosts. On the hosts' side, it may not be worth the costs to invest in counter-adaptations against manipulation; spending a season rearing a cuckoo needn't be fatal to reproductive success and might anyway be a rare event for any individual member of the host species.[9]

Cronin's proposal helps in outlining one explanation of what makes the problem of group rationality distinctive. Let us generalize Cronin's strategy thus: Whenever there is an alleged counterexample to cooperative behavior, we invoke the asymmetry of forces principle; that is, when a species, *S*, appears to be ill adapted in relation to another

manner; see L. K. DeNault and D. A. MacFarlane, "Reciprocal Altruism Between Male Vampire Bats." My account is also indebted to Helena Cronin, *The Ant and the Peacock*, 258–9, and Matt Ridley, *The Origins of Virtue*, 61–3.

[7] There is, for example, the case of the Arabian babbler (*Turdoides squamiceps*) that serves as a sentinel. It is a dangerous task, but the babbler performs that task precisely because it is dangerous; doing so serves to advertise its fitness and warn its potential predators. Despite the obvious costs it incurs, there is an evolutionary payoff for the babbler. This is not a case of cooperation (for other babblers benefit when someone else risks being a sentinel), but a self-serving case. This is the explanation offered by Amotz Zahavi in "Arabian Babblers: The Quest for Social Status in a Cooperative Breeder," "Reliability in Communication Systems and the Evolution of Altruism," and "The Theory of Signal Selection and Some of Its Implications."

[8] M. de L. Brooke and N. B. Davies, "Egg Mimicry by Cuckoos *Cuculus canorus* in Relation to Discrimination by Hosts."

[9] Helena Cronin, *The Ant and the Peacock*, 262.

species, S', S is in fact optimally adapted due to an asymmetry in the strength of the selective forces. In the present case, the selective force governing S' is stronger than that governing S. Consequently, under the circumstances, there is no opportunity for S to do any better than it is in fact doing. There is no norm or ideal that is to be specified in light of which the success of S is to be measured; nor is there room to speculate that, under the circumstances, S *could* have done better. Either there is genetic determinism, or there is not.[10] If there is genetic determinism, then the options are closed, and S could *not* have done better; if there is no genetic determinism, then under the circumstances, S *could* have done better, and the principle of the asymmetry of forces is false.

By contrast, one cannot say similar things about groups of scientists. For example, one does not say that, in seventeenth-century England, the Royal Society of London (cuckoo) was rational and so were the societies of astrologers and alchemists (hosts). Given the circumstances, the latter two societies simply did the best they could. Adopting the principle of the asymmetry of forces, in this case, would destroy the possibility of saying that a group is irrational. We could never say that it could have done better, or that it has done badly, only that it has done what it could do under the circumstances. As in the case of justice, so in group rationality: Just as one proposes an ideal in light of which various societies can be judged as being just or not – twentieth-century American society, say, is more just than was its eighteenth-century counterpart – so also in light of a normative theory of group rationality, one will be able to judge whether a group of scientists – say, the Linnean society of scientists in the nineteenth century – was (or is) rational. We are seeking a right normative theory of group rationality, not a right evolutionary explanation for why the group is what it is.

Fireflies, vampire bats, tree swallows, male olive baboons, African vervet monkeys do not commune together to discuss the question, "How ought we to organize?" Nor does one of them say to the rest, "Let's do it differently – *this* way rather *that*; perhaps then we will be more effective." The biological species, or cities, just *are*. They originate, grow, and die, each in its own distinct way; and the task of the

[10] I trust the reader will quickly surmise that I take no stand on the thorny issue of genetic determinism.

biologist, for example, is simply to explain how it in fact happened, as Cronin did in the cuckoo case. It is assumed that such evolution or process proceeds in either deterministic or stochastic fashion. As such, the biologists might use the powerful tools of game theory and evolutionary mechanism to explain those phenomena. Nevertheless, their explanation must touch empirical ground. If the end product is the same, the biologist is able to explain through computer simulation or game theory how vampire bats *could* have come to cooperate; but they do not *in fact* do so in that way, the evolutionary history was different, something would be considered missing in that account, however fascinating and illuminating it might otherwise be. Generally, the underlying assumption is that if game theory could model the end result, then arguably the evolutionary history also occurred in that way; otherwise it would be too much of a coincidence.[11]

What, then, is the lesson for us? Matt Ridley:

> But the lesson for human beings is that our frequent use of reciprocity in society may be an inevitable part of our natures: an instinct. We do not need to reason our way to the conclusion that 'one good turn deserves another', nor do we need to be taught it against our better judgements. It simply develops within us as we mature, an ineradicable predisposition, to be nurtured by teaching or not as the case may be. And why? Because natural selection has chosen it to enable us to get more from social living.[12]

First, if reciprocity in society is simply an ineluctable result of our natural instinct, then it is not something that can be argued about and rationally recommended. As Kant says, one cannot command a moral agent that he ought to pursue his happiness, because that is what the agent does by natural disposition; by contrast, one can command a

[11] But sometimes one wants only to establish the possibility of what might hitherto have been regarded as, evolutionarily speaking, improbable, if not impossible. See Richard Dawkins, *Climbing Mount Improbable*, especially 160–7, on a computer simulation, devised without making any optimistic assumptions, of how the eye might have evolved. His work is indebted to that of two Swedish biologists, Dan Nilsson and Sussane Pelger, "A Pessimistic Estimate of the Time Required for an Eye to Evolve." The opening pages of Robert Nozick's *Anarchy, State, and Utopia* offer an account of how the "Minimal State" could have come into existence. Not that this was not quite fascinating fiction, but fiction it was: Nozick was attempting to show how we could get from where we were *not* to where we *ought to be*. It would have been nice had Nozick also told us how we could get from where we *are* to where we *ought to be*. For some trenchant observations on Nozick's state-of-nature explanations, see Bernard Williams, *Truth and Truthfulness*, especially 31–2.

[12] Ridley, *The Origins of Virtue*, 65–6.

moral agent to concern himself with the happiness of others, because that concern is not a part of the repertoire of his natural inclination. The important trick, of course, is to show (what Hobbes tried to show in the context of social theory) why it is more rational to cooperate than not. But the second point is of far greater significance. Even taking for granted that each individual scientist has an instinct to cooperate, it would not follow that it is by any means obvious *how* they should cooperate. The philosophers of science discussed in this book, at any rate, are one and all interested in the latter question rather than in showing how cooperation does in fact evolve. In any event, the answer to the first question scarcely offers any significant clues to answering the second.

We are interested not merely in *how* things happen – especially to us human beings – but in how we *ought* to let things happen. If the behavior of human beings were in principle explainable by either deterministic or stochastic laws, then such a question, in my view, would be pointless. Humans, like vampire bats, could not do other than what they are genetically programmed to do, taking into consideration whatever specific environment they happen to be in. Consider, then, scientific societies or groups. One might take an empirical approach and ask, as an historian might, about the genesis, growth, and decay of scientific groups such as the American Crystallographic Association, the American Astronomical Society, or the Canadian Society of Exploration Geophysicists. Or one might ask the question, "How ought a scientific group to organize?" This question is governed by several presuppositions, among which are the following: Groups are defective as they are; groups can be different than they are; groups can be consciously directed along a desired path; and, when the new state is achieved, groups will be at least a bit more rational than they were before. Note that each presupposition is in turn governed by an ineluctable normative assumption that enables us to gauge defect, understand and channel effort to eliminate that defect, and measure progress.

II. Not a Game Theory Problem

Let me draw upon that marvelous little book *Evolution of the Social Contract* by Brian Skyrms. Suppose we have a cake that will soon rot (disintegrate, the referee gets it, whatever) if you and I do not share it or if our joint demand exceeds 100%. How shall we divide it? Half and half seems an eminently reasonable – just – solution, *assuming* that we have no special entitlements. One might haggle, as Indians do. Say, 80% for me, 20% for you; or if that doesn't work for you, 70% for me, 30% for you, and so on – until we reach 50% for me, 50% for you – assuming a 50% bottom line for each of us.[13]

Grant that this is what we would normally *do*, why is it what we *ought* to do? Skyrms responds that in game theory *informed rational self-interest* is the key to the answer. Assuming that each of us wants a maximum quantity of the cake, what I ought to ask for will depend on what you ask for. For obvious reasons, I will not want to ask for more or less than what is jointly available: in the former case, I will get nothing; in the latter case, my lot is not optimal. So I have an optimization problem, and so do you. A solution to our two optimization problems will put us in *equilibrium.* There will then be *equilibrium in informed rational self-interest,* a concept central to game theory. This equilibrium is called *Nash equilibrium.* Furthermore, if it is true not only that one cannot improve one's situation by deviating from the equilibrium, but also that one worsens one's situation by deviating, we have what is called a *strict Nash equilibrium.*

Now, there are several strict Nash equilibriums: I could have claimed 10% (20%, 30%, . . .) and you 90% (80%, 70%, . . .); had I asked for more, I would have got nothing; had I asked for less, I would have got less than 10% (20%, 30%, . . .). Which of these various strict Nash equilibriums constitute a *just* distribution of the cake? All, some, or only one? Thus far, knowing that we are in strict Nash equilibrium does not tell us anything about the concept of justice.[14] Let me briefly explain Skyrms's objections to the Harsanyi–Rawls' view, the better to understand the force of Skyrms's own view. In *A Theory of Justice,* John Rawls had suggested a powerful idea that he called "the veil of ignorance." This veil imposes ignorance on the bargaining individuals: They do not know their sex, economic background, class position, social status, natural assets and abilities, intelligence and strength, psychological propensities, conceptions of the good, rational plan of life,[15] *and,*

[13] The experiment by Nydegger and Owen, asking their subjects to split a dollar, confirms the idea that ordinarily most people would make a fifty-fifty split; see their paper "Two-Person Bargaining, an Experimental Test of the Nash Axioms."

[14] Just as Rawls had argued that the principle of efficiency does not tell us much about justice; see Rawls, *A Theory of Justice,* 58–62.

[15] Ibid., Chapter 3, section 24, especially 118–19.

by implication, religion, or anything else that is "arbitrary from a moral point of view."[16] These individuals reason using a maximin strategy (maximize your minimum gain) that would best serve their self-interest, but in a manner that does not violate the self-interest of others.[17] Not only that, Rawls expects that individuals will comply with these principles, and be known to comply with them, in order to ensure the stability of the society through public knowledge that each individual is pulling his weight in sustaining the society they have collectively created.[18] Any selection of principles of justice made from this vantage point is, so Rawls thought, necessarily just.

Skyrms's argument is that if an individual knows so little, either about himself or the person in charge of the distribution of resources, he has little to choose between the alternatives. Strictly speaking, knowing nothing, he should settle for individual A behind the veil of ignorance getting 90% and B getting 10%, or vice versa, and then keep his fingers crossed that when the results are announced, he will be A in the first scenario and B in the second. In short, Harsanyi and Rawls have taken away so much information from the moral agents that those agents cannot in principle establish rational preferences or choices. This has been a stock criticism of Rawls from very early on.[19] Skyrms intends to do better.

Consider his notion of *Darwinian Veil of Ignorance.*[20] Let us replace individuals in a two-person game with the strategies they adopt; individual identities being no longer important. These individuals in

[16] "Intuitively, the most obvious injustice of the system of natural liberty is that it permits distributive shares to be improperly influenced by these factors so arbitrary from a moral point of view." Ibid., 72.

[17] The maximin strategy advises that "we are to adopt the alternative the worst outcome of which is superior to the worst outcomes of the others." Ibid., 133. Is this just the kind of prudence Kant thought could not count as moral, even if in conformity with the moral? Thus, a cautionary note: Since the difference principle is fundamental to Rawls's theory of justice and there are arguments that support that principle, arguments that do not rely on the maximin strategy, Rawls is careful to note that the term "maiximin criterion" is best reserved for choice under uncertainty; see ibid., 72–3.

[18] "It is an important feature of a conception of justice that it should generate its own support." Ibid., 138. Not only must one subscribe to the system, one must do so in a manner that supports, sustains, and strengthens the system.

[19] The latest salvo has been fired by Simon Blackburn's *Ruling Passions: A Theory of Practical Reasoning,* 269–78.

[20] The argument that follows rests on Skyrms, *Evolution of the Social Contract,* 9–21.

the population are randomly paired; the cake represents a quantity of Darwinian fitness defined in terms of offspring viability; and the individuals' strategies are built in, which is passed on to the offspring in accordance with the fitness they receive from their mutual bargaining in the game.

Suppose everyone in the population demanded more than 50%; they would get nothing. Suppose everyone in the population demanded anything less than 50%; each would get what he asked for, but he could increase his taking – size of the cake, fitness – by asking for 50%. If everyone in the population is asking for 75% and is paired with someone who asks for 40%, then each will get nothing. But suppose several mutants were to arise in that population, each demanding 40%. If they were paired off in the bargaining game, each would do better than the rest of the population, which would end up getting nothing. These mutants could improve their lot by raising their bid to 50%. The strategy, then, of asking 50% seems to be the *evolutionarily stable strategy*. As we saw, there were several Nash equilibria, not all of which were fair divisions; evolutionary dynamics picks *one* of the equilibria, as uniquely evolutionarily stable, and this happens to be the one that represents the fair-division equilibrium. Where other approaches had failed, evolutionary dynamics has succeeded in explaining how fair division (50/50) can come about. "Fair division," concludes Skyrms, "is thus the unique evolutionarily stable equilibrium strategy of the symmetric bargaining game."[21]

Thus far we have concerned ourselves with pure strategies in the population; but what if the population were polymorphic, marked by mixed strategies? For example, consider a population in which one-half of the population claims 2/3 of the cake and the other half claims 1/3. Call the first strategy *Greedy* and the other *Modest.* If one Greedy meets another, they get nothing; if a Greedy meets a Modest, they each get what they claimed. The Greedy's average payoff is 1/3, which is equal to the Modest's payoff. Is this a stable equilibrium? If, for whatever reason, the proportion of the Greedy population rises, they will meet each other more frequently, make nothing, and their payoff will fall below 1/3. On the other hand, if the proportion of the Greedy population falls, then they will make more than 1/3, since they will

[21] Ibid., 11.

encounter a Modest more frequently. The net result will be to keep the population at half greedy and half modest.

Let us suppose, next, that mutation introduces individuals with other strategies into the population. A *Supergreedy* mutant is one who demands more than 2/3; a *Fair-minded* mutant one who demands exactly 1/2, and a *Supermodest* mutant one who demands less than 1/3. Supergreedy mutant will go extinct, because its payoff is always zero; Supermodest mutant's payoff will be whatever it demands, but since it is always less than what Supergreedy and Fair-minded demand, it too will go extinct – except more slowly than Supergreedy. But it is interesting to note that Fair-minded will also go extinct, since its payoff will also always be less than 1/3; it will average 1/4. This polymorphism has strongly stable properties, and this spells doom for justice. There are several such strongly stable polymorphisms, some with even more pronounced inequalities.

But these polymorphic traps can be avoided if we divide the cake into several slices, permitting each individual to claim as many slices as he wants. Skyrms argues that having made the game tractable (by cutting the cake into several slices), we can assign an equal probability to every possible strategy at the start, and program a computer to run along the lines of evolutionary dynamics. What happens if you do this, says Skyrms, is that extreme strategies are eliminated and the population is taken over by the fair-minded strategy. In a run of 10,000 trials, the fair-minded strategy took over the population 61.98% of the time.[22]

Suppose, next, that we replace the random pairing of strategies with pairings that are positively correlated for like strategies. In such a case, polymorphism is not possible. Supergreedys pair off among themselves and get nothing; Supermodests pair off with each other and get what each demands; and Fair-mindeds pair off and get more than anyone else, namely, 1/2, so this constitutes the best strategy. Just as in the real world the payoffs are not perfectly granular, so there is neither perfect randomness nor perfect correlation. What interested Skyrms,

[22] The details of the experimental result are provided in ibid., 112, footnotes 23 and 24. The number of initial populations that evolve toward a fair-minded state (a state in which the fair-minded strategy dominates) depends on the granularity of the payoff. In other words, the more slices of the cake there are, the greater the number of starting populations that will end up in a fair-minded state. Of course, in reality there is a limit to how granular the payoffs can become.

however, was that even an assumption of a small positive correlation seems to have a marked effect on the dominance of the fair-minded strategy in the population.

Let us conclude this part. Other strategies, such as the strict Nash equilibrium, could not show us a way out of various results produced by non-fair-minded strategies. From the perspective of the strict Nash equilibrium, there was no possibility of distinguishing one equilibrium from another; every equilibrium seemed just. But not only does evolutionary dynamics show us a way out of this enigma, the end state it evolves into is (assuming luck in mutation) a fair-minded state. This is how justice can be seen to emerge. It leads Skyrms to a significant remark: "The evolutionary approach leaves us with one evolutionary stable pure strategy – the strategy of share and share alike. This selection of a unique equilibrium strategy is a consequence of the evolutionary process proceeding under the *Darwinian Veil of Ignorance.* In this way, the evolutionary account makes contact with, and supplements, the veil-of-ignorance theories of Harsanyi and Rawls."[23]

We saw, however, that there are stable mixed states in which different proportions of the population favor different strategies. Such a state is christened a *polymorphic pitfall.* There is a way out of it, provided there are enough random variations to usher the population into a fair-division state; alas, the population in a fair-division state can also be ejected by random variation, although the population lasts longer in the fair-division state. Finally, if the payoff is finely grained, or if there is positive correlation between like-minded strategies, a population will approximate the fair-division state, even if it does not reach it. This bodes well for justice.[24] Thus far, Skyrms.

Our task is to solve the problem of group rationality, not the problem of justice. Can the ideas that have just been sketched help us in our task? Whatever the intrinsic merit of this approach for the problem of justice – and I shall presently raise a question or two – one cannot utilize this approach, or adapt it, to solve the problem of group rationality. First, consider the issue of justice and evolutionary theory. Skyrms was

[23] Skyrms, *Evolution of the Social Contract,* 20.

[24] Skyrms concludes the opening chapter thus: "This is, perhaps, a beginning of an explanation of the origin of our concept of justice." Ibid., 21. Might he have meant the origin of one of our *principles* of justice?

very careful to say that his evolutionary account was intended not to replace a theory of justice, but only to make contact with it, supplement it. One needs a prior normative theory to indicate what is just, and then an empirical evolutionary theory to supplement it. But the relationship between the two is left unspecified. For example, shall we discard a theory of justice for which there is no evolutionary explanation available to show how the just state described by said theory of justice is a strongly stable state in equilibrium? Could different equilibria be described by different theories of justice?

Let us suppose what I shall call a *Nozickian population*: Each individual in this population has certain historical entitlements. When they are paired off in a bargain, each makes a claim to the cake he is entitled to. Suppose half the population claims 60% of the cake (they have put in more labor, time, and investment into the making of the cake) and the other half 40% of the cake (they have invested less in the venture). So long as this situation prevails, the circumstance of justice prevails; each gets *exactly* what he is entitled to. But as we saw, this is a stable state, a state one does not want to mutate out of. Then imagine a collusion on the part of 40%-ers. They argue thus: "If we bargain among ourselves, we can get up to 50%; the 60%-ers will get nothing by bargaining among themselves. They will be forced to bring down their strategy by 10%." Evolution runs over. Now equality reigns, but the circumstance of justice does not.

Problems of social justice arise precisely because resources are limited and cannot satisfy all the competing demands; take one or the other away and the problem of social justice disappears. David Hume was among the first to recognize this clearly, nor can he be improved upon:

> Let us suppose, that nature has bestowed on the human race such profuse *abundance* of all *external* conveniences, that, without any uncertainty in the event, without any care or industry on our part, every individual finds himself fully provided with whatever his most voracious appetites can want, or luxurious imagination wish or desire.... Justice, in that case being totally USELESS, would be an idle ceremonial, and could never possibly have place in the catalogue of virtues.[25]

Thus, abundant nature cures the problem of justice.

[25] David Hume, *An Enquiry Concerning the Principles of Morals*, 83.

Again; suppose, that, though the necessities of human race continue the same as at present, yet the mind is so enlarged, and so replete with friendship and generosity, that every man has the utmost tenderness for every man, and feels no more concern for his own interest than for that of his fellows: It seems evident, that the USE of justice would, in this case, be suspended by such an extensive benevolence, nor would the divisions and barriers of property and obligation have ever been thought of.[26]

Thus, restricted want cures the problem of justice.

The problem of group rationality cannot be got rid of in the way in which the problem of social justice can be done away with. To establish why not: The issue is not that once a scientific group has harvested a set of truths, its task then is to determine how it shall distribute these truths among the members of the group. What is finally harvested in a scientific group is *everyone*'s property – so to speak. In principle, a scientist who has contributed nothing to the task – of discovering the genome, or the oldest fossil, or the newest galaxy – is as much entitled to the whole discovered truth as the scientists who have spent a significant part of their lives working toward the discovery.[27] Therefore, the problem of group rationality is not the problem of just distribution. The issue is not how the cake should be divided; it is a unique kind of cake; everyone can have all of it. But *even* if we make the assumption of the commonality of the property of scientific truth, the problem of how best to divide and organize research labor remains. For the issue is *how* to structure the group in order to produce a larger cake.

Let us say that each scientist wants a maximum quantity of truth.[28] The original problem, as Skyrms envisages it, no longer arises. In

[26] Ibid., 84.

[27] Obviously, there are important social issues: How much of that discovery should be concealed from rogue states; what will be the economic impact on such free distribution of knowledge; how should the economic burdens of scientific research be justly placed; how research inputs affect acquisition of patents rights, and so on. These take us back to essentially moral, military, social, and economic matters, whence the problem of distribution of justice resurfaces. But these problems have nothing to do with the pure problem of group rationality, and may in fact presuppose the solution to the pure problem, as I shall later argue.

[28] Intuitively – in other words, awaiting further theorizing – the most truths that can be discovered in the circumstances that scientists find themselves in.

principle, I cannot ask for more than is jointly available, and what is jointly available is precisely what everyone can have. So, strictly speaking, I do not have an optimization problem, nor do you. We can put the problem of group rationality crudely thus: Let S be one way the scientific group can be structured,[29] given the various circumstances; let $S_1, S_2, S_3, \ldots, S_n$ be other ways. Our problem is to seek that group structure, given the circumstances, that will produce what each scientist wants: maximum truth. *That* will constitute our equilibrium. If we have arrived at an optimal solution – say, S_1 is that solution (for a specified circumstance) – then each scientist will be in *equilibrium in informed rational self-interest* in the scientific group. If we like, we can call this equilibrium *strict Nash equilibrium.* The interesting thing is this: S_1 is a unique equilibrium point; there are no other equilibria. When circumstances change, S_1 no longer remains an adequate solution. Then a society of scientists could improve its lot by changing its structure from S_1 to, say, S_2.

There is a way of constructing the problem of group rationality to make it resemble the problem of social justice. This is how one might do it. Each scientist in the group is interested not in every scientific truth, but only in some of them; not only that, he is interested in being the discoverer himself of those truths rather than merely and eventually their recipient.[30] So a Wilson may be interested in biology, a Chandrashekhar in astrophysics, a Lévi-Strauss in anthropology, while a Shils may engage himself in sociology. Each wants the scientific group to be so structured that it will produce more truths in his particular field than will be yielded by any other alternative structuring of the group. One might conclude that what would be a strict Nash equilibrium for a Shils (when his group is in S_1) would not be the equilibrium for a Chandrashekhar (when his group is in S_2), and vice versa. And so the

[29] The fascinating problem of *what* defines the structure of a scientific group is left unaddressed for now. Since I am now delving into other concerns, I set this one aside; I shall return to it later. Let it suffice to say that philosophers who work on the problem of group rationality as discussed in these pages define the structure of the group in terms of the *theories* scientists are working on; I define it in the first instance in terms of *method.* The point I am making should hold on either assumption.

[30] I shall treat this latter issue – of each individual wanting to be the discoverer himself – more fully in Chapter 5, where I show that a theory of group rationality that claims to preserve and protect this claims leads to bizarre results.

problem of group rationality is not any different from the problem of social justice after all.

Can this challenge be met in principle? Not to my way of thinking. There is *no* answer to the problem I have just constructed. It asks us to compare apples and oranges. Unless there is a way of comparing the truths of sociology to the truths of meteorology, or the truths about quarks to the truths about economic development, I simply do not see how a solution to this problem is possible even in principle. In the absence of that solution, it would be impossible to judge what makes, say, S_2 better than S_1. The course I shall follow is a bit more prudent. I shall assume that the problem of group rationality is to be solved for a specific discipline; I shall also assume that every scientist in a subdiscipline of that discipline will regard its want of maximizing truth satisfied if S_2 is more productive than S_1.[31]

Finally, unlike in the case of justice, with its concept of fairness as 50/50 division (all things being equal), we have *no* clear, sharp, well-honed, and settled notion of group rationality. Consider, for example, the skeptical view of Paul Feyerabend.[32] If Feyerabend is right, then a rational scientific society is one in which there is an endless proliferation of methods and scientific theories; if Karl Popper, Imre Lakatos, and others are right, then a rational society is one in which there is one dominant method (either the methodology of falsification or the methodology of research programs) as well as proliferation of some viable theories. It would be well-nigh impossible to provide an evolutionary account without knowing *what* that account must explain as rational. This is clearly the case where any purported analogy between justice and science collapses, and with it the possibility of an evolutionary explanation or an explanation via game theory.

III. Ramsey and Group Rationality

Let us wend our way along a different path. Examining Frank Plumpton Ramsey's classical approach,[33] as outlined in Simon Blackburn's

[31] There are some fascinating issues here that must eventually be addressed. For now I am portraying in broad strokes the lay of the land.

[32] For details, see Chapter 4 of this volume.

[33] See Ramsey, "Truth and Probability."

Ruling Passions: A Theory of Practical Reasoning, Chapter 6, "Game Theory and Rational Choice," we find that our plea for not using game theory for solving the problem of group rationality is made perspicuous.[34] My aim here is not to examine Blackburn's claims, or to determine how persuasive they are in the field of moral philosophy (not being a Humean, I find myself unpersuaded), but simply to accept Blackburn's analysis and see what consequences flow from applying that analysis to the problem of group rationality. The consequences, we shall find, are unpalatable. To that end, I shall adapt Blackburn's Blackmail argument to the problem of group rationality and show that, at best, on such an analysis of the problem, the solution to the problem of group rationality is presupposed.

It is best to begin with the notion of *homo economicus*, economic man. As a rational actor, economic man is wedded to the fundamental principle of rationality, namely, that one should always act on the principle of maximizing one's own expected utility (given background beliefs). A rational actor is a basic unit in the society; when he has to make a choice, he avails himself of the necessary information, not only about the various alternatives at his disposal, but also about the consequences that will result from adopting each of the alternatives; and finally, when he chooses among the alternatives, he makes that choice that will enhance his own self-interested goals, ensuring that the preferences reflected in his choices are stable and consistent.

Blackburn then states that the fundamental principle of rationality – that one should always act on the principle of maximizing one's own expected utility – can either be an empirical truth, a normative truth, or an analytic or definitional truth. Devoting the better part of Chapter 5 to the topic, Blackburn argues that the fundamental principle of rationality can be neither an empirical truth nor a normative truth. Because Blackburn's reasons for those claims are not germane

[34] In 1962, in "The Use and Misuse of Game Theory," Anatol Rapoport had offered, albeit briefly, reservations, and prospects for a solution, similar to those offered by Blackburn. See also Anatol Rapoport, *Fights, Games, and Debates*, and the counter-argument proposed by Morton D. Davis in *Game Theory: A Nontechnical Introduction*. Ken Binmore, *Game Theory and the Social Contract*, volume 1, offers an advanced, elegant approach, but with much the same Humean tilt. William Poundstone's *Prisoner's Dilemma* was also helpful.

to our inquiry, I shall accept his arguments and move on. Chapter 6 is entirely given up to underwriting Blackburn's Ramseian claim that the fundamental principle of rationality is an analytic or definitional truth.

It is quite important to be clear about the relationship among choices, preferences, and utilities. Choice is the primitive term; to understand choices, we hypothesize preferences; to understand preferences, we hypothesize utilities. In short, utilities are the ultimate theoretical constructs in terms of which we understand preferences; and preferences are the penultimate theoretical constructs in terms of which we understand choices, there being relatively no ambiguity about what is choice or regarded as chosen. When a person makes a choice, he acts as if he were armed by a preference relation that orders the options he has available. To understand the constraint on the preference relation, we have to understand the notion that *a* is weakly preferred to *b*: *a* is weakly preferred to *b* if and only if the expected utility of *a* is at least as great as that of *b*. A person's preference relation must then satisfy the following two conditions: For any two choices *a* and *b*, either *a* is weakly preferred to *b*, or *b* is weakly preferred to *a* (the totality condition); and for any three choices *a*, *b*, and *c*, if *a* is weakly preferred to *b* and *b* is weakly preferred to *c*, then *a* is weakly preferred to *c* (the transitivity condition).

Anyone who satisfies the foregoing two conditions is an *eligible* agent. "It is extremely important," however, says Blackburn, "not to confuse the issue by calling them rational, as is frequently done, because this perpetuates the illusion that we are talking about special sorts of person, or giving *recommendations* in the name of reason. Whereas all we can say so far is that an ineligible person would simply be someone who cannot be interpreted in terms of utilities, so far as the set of options in play is concerned."[35] Short of forfeiting eligibility, anyone can be thus interpreted; just as short of being illogical, the basic principles of logic provide a grid in terms of which any agent's inferences can be interpreted.

With that as the backdrop, let us turn to the example of Blackmail, adapting it from Blackburn's version of it. There are two scientists, Adam and Eve. They both need resources to conduct their scientific

[35] Blackburn, *Ruling Passions*, 164.

work, and their work is mutually supported. Now, Eve has committed a professional indiscretion (minor cheating, failing to give proper credit, slightly fudging her data, whatever), and Adam knows of that indiscretion. If Adam does not squeal on Eve, he gets $1 and Eve gets $2. If he threatens to blackmail Eve and Eve surrenders, she gives up $1, Adam getting $2 and she $1. If she not only does not surrender but also threatens to reveal Adam as a blackmailer, then each ends up with nothing. Using the common terms, we shall call the aggressive option "hawkish," the conciliatory option "dovish." What should these two scientists do?[36]

We must now reckon Blackburn's cardinal distinction between an empirical and a theoretical game, a distinction rarely made in some versions of game theory. The empirical game is defined in empirical terms: For example, if one person chooses *X* and the other chooses *Y*, one gets so many years of prison, or so many dollars, or so much relief from pain and suffering. "But describing a problem in these terms," says Blackburn, "is not describing it in terms in which an agent necessarily sees his or her situation."[37] How the agent actually see his situation will determine the choice he makes. Thus, knowing full well that he will end up in prison, a person might tell the truth just the same; how many years he will spend in prison is given to him in the empirical game. What that game does not give him is what is he going to do with that information: whether he is going to risk going to prison or do something that will avoid that consequence. This is given to him by the theoretical game, when the game is *interpreted.*

Using this distinction in our example of the two scientists, Adam and Eve, we say that the empirical game is the payoff in dollars; but that still will not tell us anything about their respective theoretical games. And it is *that* game that ultimately matters, because it is the theoretical game that will dictate the final choice. As a first step, let us say that their empirical game coincides with their interpreted game; the only thing that moves them is dollar values or years in prison. Grant, further, that each of them knows that that is true of the other. Then, essentially, there is no dilemma; we have arrived at the inevitable. Adam knows that Eve will surrender, since $1 is better than nothing for her; Eve

[36] Ibid., 171–2.
[37] Ibid., 168.

knows that Adam will blackmail, since $2 is better than $1 for him. Their respective choices are governed by their preferences, and their preferences are determined by their utilities.

But, suppose that Eve is made of sterner stuff, unwilling to brook blackmail. The empirical game tells us that if she does not surrender, she will end up with nothing. Now the interpreted game kicks in. For if Eve chooses not to surrender, we would perforce have to read higher utility into that choice, indicating a higher preference for defying Adam than for surrendering to him. Consequently, we would have to conclude that for Eve the empirical game does not define her theoretical game; for Eve, the two games are different. Says Blackburn:

> In a real-life situation, Adam may know the empirical game. But he is very unlikely to know the theoretical game: in fact nobody is likely to know it until after the actions have been taken, and revealed the agent's concerns.... The game theorist can *only* say: be someone, or don't be someone, who is modelled as being in this or that theoretical game. What she *cannot* say is: once you are in this or that empirical game, play this or that way.[38]

And then:

> At this a game theorist may well complain that we have illegitimately introduced concerns that are not represented in the game's pay-off structure. In other words, we are (arbitrarily from his point of view) specifying that the theoretical game is not like the empirical game. But if he stringently insists that the numbers do represent expected utilities, then there is nothing paradoxical or surprising in the [foregoing] result.... Remember that... nobody ever chooses in such a way that their expected utility is diminished by the choice. It is a matter of definition that people act so as to maximize expected utility.[39]

I now draw the conclusion I wish to impress upon the reader.[40] Given this account of utility, preference, and choice, the distinction

[38] Ibid., 172.

[39] Ibid., 174.

[40] The example on which my conclusion about group rationality is based closely tracks, admittedly, the simple example from Blackburn. Of course, one could have offered an example mirroring a more complex source of conflict between Adam and Eve: For example, it could have been a conflict over a scientific theory, its metaphysical underpinnings, method, ways of solving a problem, evidence evaluation, or theory selection. Furthermore, the positions of Adam and Eve could have been shown to be just as flexible or entrenched, and these positions could have been permuted, as in our simple example. But nothing much would have been added to the argument except complexity.

between an empirical and a theoretical game, and Blackburn's claim that we had better be careful in calling the eligible agent rational, we can conclude that *no* decision made by either of the two scientists, Adam and Eve, can be deemed irrational so long as the totality and the transitivity conditions are observed. But these conditions can obtain in every which way. Consider: If they both cooperate (are dovish), they satisfy the two conditions and they are not irrational. If one of them does not cooperate, he (or she) satisfies the two conditions, too – we just have to make clear that the interpreted game is different from the empirical one; the person failing to cooperate is maximizing her expected utility (given background beliefs). If neither of them cooperates – they both act hawkishly – even then both of them have satisfied the two conditions and successfully managed to maximize their respective expected utilities. That they receive nothing is no indicator that they are irrational, because that is only the empirical game, not the actual theoretical game. No other alternative remains.

Assume a total number of scientists N. If they were to follow the anarchic method of Kropotkin or Feyerabend, they would be deemed to satisfy the two conditions, since if they chose to cooperate in their scientific task in this anarchic manner, it would clearly indicate that their choice is underpinned by what they believe would maximize their expected utility. On the other hand, if m scientists chose to adopt one method, say, the Popperian method, and another group of scientists, $N - m$, adopted another method, say, the Kuhnian method, around which to organize the rational structure of their society, that too would satisfy the two conditions, clearly indicating that *that* is how they deem their expected utility to be maximized. Therefore, *no* solution satisfying the two conditions can be deemed irrational.[41]

[41] That this is exactly right can be seen from what a distinguished Ramesite says: "A striking example is the common assumption that 'Truth and Probability', like modern Bayesian decision theory (e.g., Jeffrey's *The Logic of Decision* 1965), tells us to 'act in the way we think most likely to realize the objects of our desires' whether or not those thoughts and desires are either reasonable or right. It does no such thing: it claims only that the psychological theory which says that we do in fact act in this way is 'a useful approximation to the truth'. Ramsey's claim is purely descriptive, the appearance of prescription arising only from misreading his first-person 'should' in sentences like 'the more confident I am ... the less distance I should be willing to go ... to check my opinion' as 'ought to' instead of 'would'." D. H. Mellor, "Introduction," xviii.

One can see this unsettling consequence from a slightly different stance. In trying to solve the problem of group rationality, we are trying to come up with a *normative* theory. A normative theory is one that recommends how one *ought* to act, behave, or structure. One can transgress a norm and thereby make a mistake – one very often does – and one is thereby deemed blameworthy. On the other hand, Ramsey's theory, at least in its Blackburnian version, is not, cannot possibly be, normative, because one cannot in principle make mistakes; it is explicitly argued to be definitionally or analytically true. There is, therefore, *no* possibility of enlisting Ramsey's theory in the service of a theory of group rationality. Whatever the decision of the scientists (with respect to theories) and whatever the structure of their group, so long as the two conditions (the totality and the transitivity conditions) are satisfied, it is rational.[42]

There is one more path to traverse before we turn to examine some of the classical theories of group rationality, and that is to see how some of the ideas in welfare economics and social choice theory might hold promise for what we are about. There is no better way to do that than to draw on the work of Amartya Kumar Sen; it is to that task that I turn next.

[42] Bernard Williams says, "Science is, in game-theoretical terms, not a two-party game: what confronts the inquirer is not a rival will, and that is a key to a sense of freedom that it can offer." *Truth and Truthfulness*, 144–5. What follows for a few pages thereafter is useful and interesting. But while Williams is entirely correct in saying what he does, he does not go nearly far enough. Important as the sense of that freedom is, the question that the problem of group rationality brings to the fore is what the scientists are to do collectively with that freedom – how they are to organize – to pursue the truth. The answer to that question scarcely follows from adverting to the importance of the scientists' freedom; indeed, the question is not even raised.

3

The Problem Explored

Sen's Way

Let me here weave in some of the powerful ideas of Amartya Kumar Sen. Among his works I have principally relied on are *Rationality and Freedom* (2002), *The Standard of Living* (1988), *On Ethics and Economics* (1987), and *Utilitarianism and Beyond* (1982). One cannot map a hectare, a reader might think, by squinting at an acre; and he would be right. As it is, the task I set myself is considerably more modest. I am not even attempting an analysis of a few key notions in Sen's philosophy, let alone undertaking to provide a large-scale, systematic account of his dominant views. Rather, I am more interested in an eclectic reading of some of his notions – with a certain sense of adventure, I admit – to see if anything might emerge that will be of interest to those of us working on the problem of group rationality. If I have succeeded, then perhaps Sen might also be read for reasons other than those for which he is so widely read.

Providing, then, only the barest outline of some of Sen's key notions and arguments, and duly adapting them, I sketch a set of issues relating to the problem of group rationality. My hope is not only to unravel something intriguing about the realm of group rationality; it is also to show how some of the concerns in this realm may well diverge from those in welfare economics, their several and useful parallels in other respects notwithstanding. I end with what I take to be a set of marvelous problems, problems I christen the Sen-problems of group rationality.

In section I, I provide as a backdrop some of the principal ideas that inform Sen's notion of rationality. In section II, I adumbrate several other such notions, also in Sen's repertoire, and show how they might be connected to the notion of what I call *scientific welfare*, a notion entirely rooted in the Senian scheme. In section III, based on the foregoing, I present what I think are some rather interesting versions of the problem of group rationality. This alone should demonstrate that there is much work ahead of us, significant work of a preliminary nature, work that should be centered on how best to understand and analyze the *problem* of group rationality in the first place before we start offering solutions to the problem. The final, brief section IV defines the problem of group rationality that guides the remainder of this book.

I. Consistency, Ordering, and Rationality

In standard economic theory, an economic agent is considered to be rational if he is *consistent* and *maximizes self-interest*. Sen has argued that the consistency condition is hardly sufficient for rationality[1] – under certain severe conditions, perhaps not even necessary – and that the condition of maximizing self-interest (by failing to heed the range and complexity of an economic agent's motivations) has deprived

[1] For a detailed, technical account, see Amartya Sen, *Rationality and Freedom*, Chapter 3, "Internal Consistency of Choice." In his paper, "Is the Idea of Purely Internal Consistency of Choice Bizarre?" Sen perhaps carries the matter a bit too far. He argues that the consistency condition can be challenged thus: He, Sen, has a desire, an end, to appear as the inscrutable Oriental, and so he baffles his interlocutors by being contradictory; see 21–2. As a challenge to the principle of consistency, this will not do. If Sen is inconsistent at the first-order and his actions appear to be well-designed, or rationally motivated, it is because implicitly he has risen to the second-order, where his end is to be consistently inconsistent. If his second-order ends are also inconsistent, then I submit there is no action Sen could, in principle, carry out that would conjointly satisfy his second-order ends. Thus, at the second-order he would inconsistently have an end to appear inconsistent as well as to appear consistent: No matter what he does, he will frustrate one of his second-order ends. This, thus far, restores the claim that Sen is keen to deny, namely, that "some condition of internal consistency of choice can be specified *without* knowing what the motivation behind the choice is, and in the insistence that this condition must apply *no matter* what the motivation is," 21. Fortunately, Sen offers this example only as a "contrived counter-example" and, having offered it as a preliminary, moves far beyond it. Compare Chapter 4, this volume, footnote 18 and the associated text.

economic theory of its full potential to explain economic phenomena, thereby restricting its testability.[2] Now, unlike empirical economic theory, a theory of group rationality, like ethics, is a normative discipline, and hence not primarily in the business of explaining or making predictions; its "testing" is done on other grounds. Perhaps the notion of maximizing self-interest may still be relevant for group rationality.

Sen wants to widen the scope of rationality in economics no less than in ethics. One must, of course, distinguish between the principles of institutional public policy (parallel to the principles governing social justice) and the principles of an economic agent's individual decisions (parallel to the principles governing an individual moral agent's decisions). When there are several options, one option may be more valuable with respect to one dimension, another option more valuable along a different dimension. There are three approaches of rational choice in such situations. The first approach: When an economic agent has several options before him, he has to rank order his choices, along some value system, resulting in *complete ordering*, determining what *on balance* is the economic agent's best choice. This approach demands that conflicts between options be resolved *before* the final decision is made.[3] The second approach allows two or more options to be left unordered, resulting in *partial ordering*, with the final decision based on such ordering.[4] Both of these approaches insist on consistency. The third approach – not one that most philosophers or economists

[2] Hilary Putnam distinguishes between "maximizing genuine long-term self-interest" and "maximizing mere short-term self-interest." Putnam then asserts that the former is not usually identified with self-interest; see Putnam, *The Collapse of the Fact/Value Dichotomy and Other Essays*, 52. I should contend that Sen's arguments apply to the former as well. In other words, Sen would argue that an agent can act in the interest of someone else, and the same act cannot be understood as in the agent's genuine long-term self-interest.

[3] In some cases, neither consistency nor complete ordering may be necessary to determine the *best* element in a given set of choices; see Sen, *On Ethics and Economics*, 68, footnote 10.

[4] The key notions here are that of "*x is weakly dominating y*" if and only if *x* is no lower than *y* with respect to each dimension, and, "*x is strictly dominating y*" if and only *x* is higher with respect to some dimension than *y*, but not lower than *y* on any other dimension. Thus, a partial ordering may well be both incomplete *and* consistent. In the problem of group rationality, what would be the various dimensions (other than truth) along which a scientist would rank order his preferences? See discussion of *dominance reasoning* in Sen, *On Ethics and Economics*, 66, and Amartya Sen and Bernard Williams, *Utilitarianism and Beyond*, 17.

endorse – allows for inconsistency by permitting the admission of "the superiority of one alternative over the other and the converse." This last approach, cautions Sen, is at least not obviously mistaken because both the feasibility and the necessity of consistency has to be justified, and arguably they are not yet justified.[5]

In the case of institutional public policy, Sen recommends the first approach. Here is his reason:

> Indeed, needs of policy do require that something or other must be ultimately done; if only nothing, which is one such something. However, it does not follow – and this is the important point to get across – that there must be *adequate reason* for choosing one course rather than another. Incompleteness or overcompleteness in overall judgements might well be a damned nuisance for decisions, but the *need* for a decision does not, on its own, *resolve* the conflict. This implies that sometimes even institutional public decisions may have to be taken on the basis of partial justification.[6]

Is this irrational? No, it is not: "There is, I believe," says Sen, "no departure from rational choice in this acceptance."[7] Now, a philosopher might argue that there is a nice parallel, on Sen's own grounds, between institutional public policy decisions and an individual economic agent's decisions. First, the same argument used to justify the adoption of the first approach to public policy would surely be recommended in the case of an individual economic agent's decision. An individual, no less than public policy, requires that something ultimately be done, even if the decision is to do nothing. Second, neither in the individual's case nor in the case of public policy does it follow that the choice of a course of action be adequately justified: In neither case will the "need for a decision . . . , on its own, *resolve* the conflict." It follows that sometimes the decision of an individual economic agent no less than institutional public policy may have justification based on only partial ordering. Third, if "whatever value there might be in acknowledging the 'richness' of inconsistency arising from conflicts of principles, is typically personal to the individual involved in the

[5] Sen, *On Ethics and Economics*, 65–8; see also Sen and Williams, *Utilitarianism and Beyond*, 16–18. See also footnote 1.
[6] Sen, *On Ethics and Economics*, 67.
[7] Ibid.

conflict,"[8] might one not claim that a society as a whole, too, may find *itself* deeply embedded in situations where principles that work well in most circumstances are jointly unsatisfiable in certain unique – rich, if you will – circumstances? Fourth, and finally, if one rightly distinguishes, as Sen does, the need for a decision in conflicted circumstances from what rationally resolves the conflict, then it is arguable that such a distinction is necessary not only in the case of public policy, but also in the case of the decision of an individual economic agent (as in Agamemnon's case?).[9]

How would the foregoing translate into matters pertaining to group rationality? *Very* briefly, as follows. Let me begin by making a large assumption: Each scientist is interested in truth, and not in something that has nothing to do with truth; and so scientists are not divided over whether theories should be empirically adequate, metaphysically true, or anti-realistically true, or bearers of a Peircean notion of truth, and so on.[10] This is a powerful, simplifying proviso, designed to unveil what is complex and interesting in the problem at hand. In ethics, Sen has quite justly argued for jettisoning the *homogeneous product*, such as satisfaction, happiness, or utility; and such an ejection would naturally lead to the acceptance of a diversity of values, values that ultimately govern the motivation, actions, and decisions of moral agents. The homogeneous product – truth – that each scientist is supposed to pursue, and that is peddled here, is not as obviously suspect as a homogeneous product proffered in ethics. My purpose in settling on the notion of truth *simplicter* is to show that even under this highly restricted condition, we will run into an interesting problem. Dropping that condition would make the problem virtually intractable.

[8] Ibid.

[9] Ibid., 66.

[10] I chose *truth*, but I could just as well have chosen *empiricaly adequacy* as what unites the group of scientists in their aim. Using Sen's distinction between *competitive plurality* and *constitutive plurality* (see Amartya Sen, *The Standard of Living*, 2–3), one could say that by restricting the scientific goal of the group to a single goal – namely, truth – I have eliminated competitive plurality. Perhaps constitutive plurality would still remain, each scientist defining truth in his own distinctive philosophical way. The latter possibility would still leave open the rather interesting question of what is the common factor – running through the various definitions of truth – that knits the group into a cohesive whole. It is not too difficult to see that allowing competitive plurality with respect to the goals of science would make the problem of group rationality that much more difficult to solve.

To proceed, let each scientist make a choice about, say, which scientific domain to work in, which scientific problems to work on, or which scientific theories to accept for further research and development. Even restricting his domain to one scientific field (just as an aside for now: *how* is it ensured that the distribution of scientific labor in terms of numbers of scientists is just right?)[11] each scientist will have a plethora of choices of problems and theories. In making his decision, each scientist must attempt both to be *consistent* and to *maximize his self-interest.* There are three possibilities. His decision might be based on a complete ordering, a partial ordering, or it might allow an inconsistency (under certain circumstances). Now, an inconsistency, as we saw Sen argue, may not in all instances be irrational. But for our purposes, I shall assume, to see if anything of interest emerges, that no scientist finds himself in the position of having to make an inconsistent choice, and in fact that no need arises for him to make a choice based on only partial ordering.

II. Other Notions of Rationality and Scientific Welfare

Intimately linked to these several notions of rationality are a host of others. Let me simply adapt them for the purposes of the problem of group rationality (without presenting their correlative in economics). Not only should there be *instrumental rationality,* there should also be *correspondence rationality* – where what is chosen corresponds with what is aimed at – which is to be supplemented by *reflection rationality* – where reflection on what is wanted, valued, or aimed at is constrained by rationality requirements. Sen rightly regards correspondence rationality as at the very least a necessary condition of rationality, because a person who merely consistently picks the opposite of what he wants can scarcely be deemed rational; but correspondence rationality may not be a sufficient condition unless complemented by reflection rationality, which critically assesses and endorses what it is that one wants.[12]

[11] Crudely, if scientific domains can be objectively rank ordered in accordance with their value and importance – revealing a hierarchy among the sciences – and assuming that every scientist agrees with that ranking and hierarchy, then just *how* is it ensured that the number of scientists working in each field is directly proportional, or even roughly proportional, to each field's rank and need?

[12] Sen, *On Ethics and Economics,* 12–14 and footnote 9.

Let us take these three notions of rationality in reverse order. First, by stipulation, we have neutralized what might have been potentially problematic. Thus, reflection rationality assures each scientist that his aim should be the pursuit of truth; had we not so stipulated, we could have faced an impasse between some scientists in the group who might maintain that the aim of science should be merely empirical adequacy, as opposed to some other scientists who might maintain that its aim should be truth in the sense of metaphysical realism (with various other subgroups of scientists preferring still other options). Allowing for dissonance regarding the aim of science would, of course, make the problem of group rationality exceedingly difficult.[13]

Second, given that the aim of science is the pursuit of truth, the scientists' choice must reflect correspondence rationality, namely, that what is chosen corresponds with what is aimed at. In short, given the field or domain he wishes to work in and the theory he has chosen to pursue, the choice or choices of the scientist must correspond with his aim to get at the truth in that domain.

Third, instrumental rationality, says Sen, which does not answer to a critical assessment of what it is that a scientist is trying to achieve, is quite inadequate. It follows that a scientist who has satisfied the foregoing two conditions will have automatically satisfied this third condition.

And then there is the following (which, I have directly adapted):

Self-centered scientific welfare: A scientist's welfare depends only on his or her own consumption of scientific theory (and in particular, it does not involve any sympathy or antipathy toward fellow scientists). Consumption is here to be understood as finding satisfaction in understanding, developing, proposing, and testing a scientific theory.

Self-scientific welfare goal: A scientist's goal is to maximize his or her own welfare, and – given uncertainty – the probability-weighted expected value of that welfare (and in particular, it does not involve directly attaching importance to the welfare of other scientists).[14]

Self-scientific-goal choice: Each act of choice of a scientist (in relation to domain, problem, and theory) is guided immediately by the pursuit of his

[13] I try to grapple with this problem in the next volume.

[14] And, with Rawls, I too shall make the assumption that the scientists are free of envy; see Rawls, *A Theory of Justice*, 124 and section 80. If this appears as an innocent enough assumption, then see Chapter 5 of this volume, footnotes 31 and 32.

or her own goal (and in particular, it is not restrained or adapted by the recognition of the mutual interdependence of respective successes, given other scientists' pursuit of their goals).[15]

Let us concentrate on the first notion, *self-centered scientific welfare*, since the other two notions are parasitic upon it: Choice is dependent on goal, and goal is dependent on welfare. The notion of scientific welfare also deserves scrutiny because it may well prove intriguing for our purposes. But first, some backdrop. "A functioning," says Sen, "is an achievement, whereas a capability is the ability to achieve."[16] Functionings are closely connected to actual living conditions; by contrast, capabilities are notions of freedom, connected to the notion of real opportunities that one might have in one's life. It might appear that we need to focus only on functionings, but that would be a mistake. Capabilities are important in assessing the quality of one's life. If a scientist has a choice among scientific lifestyles *A*, *B*, *C*, and *D*, and he chooses *D*, then, if he is deprived of the first three opportunities, but can still pursue lifestyle *D*, his functioning has not altered, but presumably his freedom has. Next, consider two individuals who have an opportunity to share a cake as opposed to tossing a coin for it, winner take all. Suppose the consequences of not having a slice of the cake are dire. Would the loser in this venture consider his freedom restricted if he did not have the opportunity to toss for the cake?[17]

Say, then, that each scientist may be seen as having enough opportunities – they are free in the Senian sense (or whatever is the Adam Smithian equivalent in the sciences of appearing in public with linen shirt and leather shoes, and thus appearing without shame) – in the context of his or her actual scientific life.[18] There is no restriction on

[15] Sen, *On Ethics and Economics*, 80; see also Sen, *Rationality and Freedom*, 33–7.

[16] Sen, *The Standard of Living*, 36.

[17] This simple example has telling consequences; I borrowed it from Ravi Kanbur, "The Standard of Living: Uncertainty, Inequality, and Opportunity"; see especially 60–6.

[18] Amartya Sen, *Development as Freedom*, 73–4.

Recently, nothing has demolished more effectively and systematically the view that Adam Smith's *Wealth of Nations* is an "Employer's Gospel" than Emma Rothschild's *Economic Sentiments: Adam Smith, Condorcet, and the Enlightenment.* Smith (as well as Condorcet), Rothschild argues, was as interested in equity as he was in efficiency; indeed, the twin notions are entangled, interlaced. Rothschild's primary interest is in the history of economic thought of the harrowed eighteenth century; my interest, on the other hand, keeping in mind the views of Sen, guided me to cull from her book what may inform, and illumine, the problem of group rationality.

the kind of scientific lifestyle a scientist may adopt. Note: The fact that one is not able to move easily from the field of plant morphology to quantum field theory, say, is not the kind of restriction I have in mind. (I venture that those *kinds* of restrictions are not serious impediments in ethics and social philosophy either.)[19] Nor are scientists minded to take enormous risks of the sort outlined in the example of tossing a coin for the cake, winner take all. For example, no scientist will toss a coin to decide whether the theories in his own domain of micropaleontology should be wholly and entirely researched and developed over theories in a domain of neurobiology in which he has no interest (as per the notion of self-centered scientific welfare).

Perhaps this is the kind of atomism Charles Taylor was concerned to criticize when he said that "the free individual cannot be concerned purely with his individual choices to the neglect of the matrix in which such choices can be open or closed, rich or meager" and that "it is important to him that certain activities and institutions flourish in

Thus, consider: Both Smith and Condorcet were concerned to promote individual justice and liberty, whether they were discussing the problems of the corn trade, famines, lending at interest, the apprenticeship system, the English Poor Laws, environmental conflict, or education, and not just efficiency. Essentially, their argument was not only that the promotion of freedom – a " long-drawn-out" process – would, in time, bring about economic prosperity – and such freedom was a person's inalienable right – but also that economic prosperity was necessary for any meaningful liberty. This entailed that it was not part of the governmental task to ensure happiness so much as it was to provide welfare. Benjamin Constant, voicing Condorcet, said of the government: "Let it limit itself to being just; we will take it upon ourselves to be happy." To be sure, real procedures are often imperfect for a variety of reasons; but again, the governmental task is to implement policies that would correct these procedures rather than to influence outcomes. Indeed, where said welfare was dependent on enlightenment, Condorcet was even unwilling to externally impose enlightenment. See Rothschild, *Economic Sentiments*, especially 19–20, 38, 61–4, 68, 77, 101, 103–4, 161–2, 166–9, 186, 192, 200–1, 215, and 219–20.

The task of group rationality, by contrast, is not to show how to bring about justice and liberty, which in turn will promote scientific welfare. The freedom of the scientists is *assumed*, as is equity. Our task begins thereafter. The task is to show (*a*) *which* of the group's possible states will make for its rationality (leaving undefined for now the notion of the state); (*b*) *how*, given freedom and equity, scientists will rationally come to agree that such a state of their group is rational; and (*c*) *how* the scientists would set about producing that state. Problems (*a*) and (*b*) are by far the more interesting; I shall have little to say about (*c*) in this book.

[19] John Rawls, for example, is concerned primarily with whether opportunities are available to a moral agent rather than with what, if anything, the agent accomplishes when given those opportunities; see John Rawls, "Social Unity and Primary Goods."

society."[20] The view of the problem of group rationality presented here escapes that Taylorian objection. The problem of group rationality is not to show what kind of matrix or institution a scientist would like to see flourish in his midst that would ensure the research and development of theories in his domain that maximize his self-centered scientific welfare. Rather, the problem of group rationality is to show what kind of institution scientists would *collectively* like to see flourish in their midst that, given the self-centered scientific welfare of each scientist, would ensure the optimum research and development of theories in various (perhaps conflicted) domains.

Let us stipulate that every scientist in the group satisfies each of the foregoing requirements of rationality: He is consistent; he makes choices that are governed by his goal; his goal is to maximize his own scientific welfare (without desiring to enhance or detract from the scientific welfare of fellow scientists), and so on. (*Question*: If we allow each scientist in the group to have a *right* to pursue whatever science he wishes to pursue, while the *interest* of the group as a whole lies in his doing something else, could something resembling the paradox of the impossibility of a Paretian Liberal follow in the field of group rationality as well?) Now, here is a point of fair significance: What differentiates the problem of group rationality from a problem in welfare economics is that in the latter when the division of labor produces n quantity of product, each economic agent has an entitlement (however determined) to a portion of n to which no one *else* is entitled; by contrast, when the scientific division of labor produces n quantity of product (in terms of theories, observational and experimental data, fresh problems, and such) *each* scientific agent has an entitlement to *every* bit or portion of n. So, in that precise sense, the problem of group rationality is not a problem of how to divide the product of scientific labor.

III. The Sen-Problems of Group Rationality

In what sense, then, is it a problem, and how, if at all, would such a problem be solved? To simplify: Suppose a group of scientists consists of two subgroups, *A* and *B*. *A* is interested in the field of ecology, *B* in the field of cosmology. (Granting, for the sake of argument, that

[20] Charles Taylor, "Atomism," 207. Quoted in Geoffrey Hawthorn, "Introduction," xiv.

interpersonal comparisons of scientific well-being are not meaningless, one might be called upon to invoke such comparisons here, even though intuitively the difficulties seem insurmountable.)[21] Whatever the result produced by each subgroup, it is made available to the other. But each subgroup has a vested interest in its own field (in which lies its self-centered scientific welfare) than in the other: A significant advancement in *A*'s field contributes to its welfare in a way in which a significant advancement in *B*'s field does not, even assuming that *A* "consumes" *B*'s theory, and vice versa.[22]

The problem of group rationality, duly generalized – let us call it *the Sen-Problem of Group Rationality* – leads to this first variant:

S_1 (*The Sen-Problem of Group Rationality*): How should *n* scientists structure their activities so that each scientist maximizes his or her own self-centered scientific welfare (where what is jointly produced is shared without remainder for consumption)?

[21] Compare Sen, *On Ethics and Economics*, 30–1.

[22] Ultimately, utilitarian considerations might be difficult to escape. John Stuart Mill rightly claimed that, in order for utilitarianism to survive crude objections against it, the notion of happiness must be made broad and inclusive. A scientist joins the scientific enterprise usually because he has a passionate interest in a particular field of science – ornithology, anthropology, weed science, ichthyology, whatever. *This* is what he delights in; therein lies his utility. Such a scientist might abstractly concede that *science* would prosper more if less attention were paid to the development of his field than to other sciences. But he cannot be thought to be motivated by that prospect alone to join in the group effort.

Or consider this: When asked, "Why do you desire *x*?"one might understandably, even acceptably, reply, "Because I value *x*." However, if when asked, "Why do you value *x*?" one replies, "Because I desire *x*," the reply seems unsatisfactory. Sen's wry comment: "This would, of course, be a good way of earning a reputation for inscrutability, but not a particularly effective way of answering the question asked." Sen, *The Standard of Living*, 10; see also Sen and Williams, *Utilitarianism and Beyond*, 6, 12–13. Without intending to generalize to other fields, one might argue that it is possible for an ornithologist to accept intellectually the twin notions that there is a hierarchy among the sciences and that ornithology does not rank very high in that hierarchy. Now, if asked, "Why, then, do you value ornithology?" the scientist might reply, "Because I just desire its study." The value of his pursuit of ornithology lies in his desires, rather than conversely. On the other hand, for the benefit of science or the group, an ornithologist might engage in a scientific task that benefits science or the group, but not himself as an ornithologist. My purpose is to argue only that if an ornithologist chooses the former alternative and not the latter, it is not obvious that he is being inscrutable. Nor is my argument predicated upon ignoring motivational complexity; see Sen, *The Standard of Living*, 13. While I am uncertain of the efficacy of Bernard Williams's famous distinction between internal and external reasons in ethics, they might play a significant role in the problem of group rationality.

"The real issue," says Sen, "is whether there is a plurality of motivations, or whether self-interest *alone* drives human beings."[23] Sen has offered unyielding criticism of the view that an agent always maximizes his self-interest (not to mention Sen's heroic attempt at debunking a widely prevalent misinterpretation of Adam Smith).[24] Denying that people always act in a self-interested way is not the same as asserting that they always act selflessly; indeed, economic transactions would be inexplicable, says Sen, if no self-interest were involved. The conclusion is neither that people never act selflessly nor that they always do, but rather that they have a medley of motives, and among those motives are ones in which other peoples' interests are heeded, if not zealously pursued and protected.[25] And allowing for such a medley of motives makes economics a more easily testable science.

[23] Sen, *On Ethics and Economics*, 19.

[24] See, among other sources, Sen, *On Ethics and Economics* especially 22–8; Sen, *The Standard of Living*, 10; and Sen, *Development as Freedom*, 122–3, 255–6, 271–2. See also Rothschild, *Economic Sentiments*, 199.

[25] Both Smith and Condorcet assumed that ordinarily individuals have a generous disposition; the "shock of opposing interests" notwithstanding, there is no reason to assume individuals to be unreasonably competitive, greedy, or unjust. Moreover, given the liberty to discuss their common problems, individual economic agents will collectively agree to do the right thing; there will be a convergence of opinions on what is fair and impartial. They will see to it – barring rare exceptions – that through liberty undue harms will not be inflicted. Furthermore, repeated discussions will correct whatever mistakes have been made in their agreement. In short, nothing will be treated as sacrosanct: In significant part, this will be due to the fact that probability will guide economic agents, and to that degree their lives will be marked by uncertainty. Any agreement they do arrive at will not be imposed externally by a government or a group of wise men and women; it will be a decision made internally by self-respecting and other-respecting economic agents. See Rothschild, *Economic Sentiments* especially, 49, 157, 187–92, 202–9, 231, 233–5, 237–8, 250.

Now, in the case of group rationality, agreements among scientists – "sagacious," constituting "a society of reasonable beings," too – will not be externally imposed or enforced. However, at the core of the argument, there will be minimal reliance on their generous disposition or Smithian sentiments, such as sympathy, even granting the optimistic assumption of "moral uniformity" among human beings. See Rothschild, *Economic Sentiments*, especially, 20, 23, 200, 190–1, 209–10, 228, 226n44. In this respect, the problem of group rationality is quite non-Humean, and to a fair degree Kantian. Or at least Kantian in this respect: Each individual scientist will require a *reason* why he should cooperate in the common scientific venture. It is not enough to say that he has a disposition to cooperate; indeed, even assuming a uniformity of such a disposition, it will leave unanswered the question "Toward *which* of the possible states ought he to be disposed to cooperate?" There is this much similarity (entirely explicable in terms Kant outlined in *What Is Enlightenment?*): Any decision that a group of scientists arrives at is not written in stone; it is always open to revision through further discussion, argument, and counterargument, especially if vetoed by a member for some reason.

This naturally prompts a worry that Sen just as naturally anticipates and states forthwith:

> (1) *The absence of a sure-fire test*: There is an inescapable latitude in this approach of deciding whether a person is being rational or not, and that might well appear to some to be quite unattractive, especially to those who like their theories and methods to be fully decisive and conveniently algorithmic.[26]

Now, philosophers of science, or methodologists, would plead not guilty to the charge that they want, or are searching for, an answer to the question "What makes a scientist rational?" that can then be used by a scientist to mechanically generate answers to questions like "Which scientific theory ought I to follow for further research?" Their fears might be expressed thus: At one end lies an algorithmic answer; at the other end, the answer is "Anything goes."[27] Clearly, and rightly, Sen has no interest in either answer. The task, then, as a methodologist sees it, is a bit more modest even than coming up with precisely *one nonalgorithmic* answer; there *is* wisdom in allowing a bit more leg-room. Consequently, the philosophical task is to show what features delimit what is to count as rational, and how those features would permit entry to more than one theory in the domain of what is rational, yet prohibit everything else from entering it.

Look at it this way. Sen asserts, "If rationality involves disciplined freedom, the non-imposition of externally dictated 'tests' is part of that freedom, just as the necessity to subject one's decisions and values to exacting scrutiny is part of the required discipline."[28] A skeptic (such as Feyerabend) might wholly agree with Sen: Sen is right, he might say, that both freedom and intense scrutiny are necessary ingredients if a person is to justifiably claim to being rational in his values, motivations, and decision making. But, the skeptic might caution, unless there are objective standards in light of which the scientist (or a moral agent) can judge, or be judged, the question "What makes a scientist rational?" will not have been adequately answered. This is especially so because Sen's "approach gives a person considerable freedom about the types of reasons that may be invoked and used, and much would depend on the person herself."[29] If no restraints are placed on this freedom,

[26] Sen, *Rationality and Freedom*, 48.
[27] Husain Sarkar, *A Theory of Method*, 149–54.
[28] Sen, *Rationality and Freedom*, 49.
[29] Ibid.

or on the types of reason that can be used, where much depends on the agent himself, the threat of skepticism would loom large on the horizon. Perhaps an adequate theory of group rationality will indicate just what are those filtering features through which an action must pass in order to earn the label "rational"; in short, such a theory will show what features draw the boundary around individual rationality.[30]

Amid raising some nice, sharp queries about Sen's notions of choice and capabilities – asking, for example, how closely the two notions are connected – Bernard Williams says, "the arguments about capability need to be taken further. One is that it is very doubtful that you can take capabilities singly in thinking about these questions: you have to think about sets of co-realisable capabilities, and about social states in which people acquire various ranges of capability."[31] Applying Williams's claim to the issue of group rationality, I want to say, first, that we might consider generalizing the idea about "sets of co-realisable capabilities" to the group as a whole. Williams correctly argues that not every capability would be realized in a social state; only sets of capabilities would be realized. Arguably, an individual would want that set realized that would maximize his welfare, and a society would want those sets of capabilities in its citizens realized that would bring the society nearer to its ideal than would the satisfaction of any other such set of capabilities. Similarly, then, not every scientific capability would be realized in a group; only certain sets of such capabilities would be realized. The task of group rationality is to determine the structure of the group of scientists wherein the several sets of scientific capabilities would lead nearer to the truth than would the satisfaction of any other such set of scientific capabilities.

Second, and briefly, while it is true that it is only in certain social states that people acquire various ranges of capabilities, the range of available social states from which an individual, or for that matter a

[30] I should think similar arguments must apply to the notion of *group* rationality. The notion should not be so narrowly circumscribed that only one theory can count as describing a group as rational, nor, of course, should it allow every theory to do so. The utter significance of individual rationality and its role in group rationality will be more extensively explored in the penultimate chapter, where we focus on the work of Hilary Putnam.

[31] Bernard Williams, "The Standard of Living: Interests and Capabilities," 99–100. For Sen's reply to Williams, see Sen, *The Standard of Living*, 107–12.

society, can choose is extremely narrow. We are generally confined to a few, paltry states. Similarly for scientists and the range of available social states in which to practice science.[32]

Simply to mark a sharp contrast to the Senian standpoint, let us formulate a utilitarian version of the problem of group rationality:

UGR (*The Utilitarian Problem of Group Rationality*): How should *n* scientists structure their activities, given that each scientist is involved in directly attaching importance to his or her own scientific welfare, and the happiness he or she derives from that welfare, in doing, testing, and producing science (and where what is jointly produced is shared without remainder for consumption)?[33]

The basic idea of utilitarianism is applied here. We engage in *sum raking*. For a given scientific social state, the happiness of each scientist in pursuing science in that state is summed to determine which scientific social state will produce maximum happiness in the pursuit of science.[34] Now, at least one ground on which utilitarianism has been held suspect is its inability to allow for a plurality of principles based on the *distinctness* of individuals and their *projects*; it turns individuals into urns into which utility or happiness is poured and robs them of their uniqueness, as of their *integrity*. This is Williams's and Sen's time-honored argument.[35] It is intriguing – certainly worth noticing – that such an argument might fail to apply to scientists looked at from the utilitarian point of view of group rationality. Each scientist is interested in truth, and his uniqueness, his project, and his integrity are closely interwoven with the pursuit of truth.[36] The scale of utility as sole value

[32] Throughout this book, I have not entertained the question, "What is the legitimate authority that could govern a scientific group?" I have set the question aside because I am not sure of the payoff to rationality or epistemology in answering it. Such a question is, of course, central in social and political philosophy; see Sen and Williams, *Utilitarianism and Beyond*, 2–3.

[33] See Chapter 7, this volume, 75, where this formulation of the problem is put to work; see also Chapter 8, 221.

[34] See Sen and Williams, *Utilitarianism and Beyond*, 3–4.

[35] The earliest appearance of Williams's argument, to my knowledge, is in his 1973 *Utilitarianism: For and Against.* Sen's stand against utilitarianism (while retaining what virtues it has), both in and out of economics, is of course well-known. Perhaps Condorcet might be deemed their precursor; see Rothschild, *Economic Sentiments*, 200–1.

[36] It is possible to have several criteria of rationality in science at one level that are subject to a higher-level criterion of rationality, such as truth. The former criteria of rationality would allow for uniqueness, project, and integrity, while the latter criterion of rationality would discriminate or evaluate those criteria; the former would allow

may be objectionable, but the scale of truth as sole value is not, at least not obviously so.[37] Utility as a common yardstick may be objectionable on the grounds that it obliterates relevant moral differences; it is difficult to see how truth as a common measure can be similarly dismissed.[38]

In pursuing a theory of group rationality, our concern is not with predictability, or with making it more testable, but rather with its normative or prescriptive aspects. Let us suppose, then, that someone – or even Sen – were to criticize us for making our scientists too self-serving.[39] He might say that if we had made a more realistic assumption and attributed a medley of motives to scientific agents, as one should in the case of economic agents, then the Sen-problem of group rationality need not appear intractable. Indeed, given the objectivity of the aim of science (truth) and hence of scientific results (rather than the messy, subjective, plural values of economic agents), conjoined with scientists' concerns being more broad-based than self-centered, the solution may be within reach.

for multiplicity, while the latter would not, or would allow only a fairly restricted set. For a possible objection to which the previous two sentences are a response, see Sen and Williams, *Utilitarianism and Beyond*, 2.

37 Ibid., 7.

38 Perhaps one might argue against it thus: One scientist may pursue science for the sake of truth, another for the sake of ameliorating health, a third to provide defense measures for his country, a fourth to put an end to poverty, and so on. If that is so, scientific individuals are driven toward different projects, and their integrity depends on achieving different things. Two replies are in order: (*a*) In these instances, the motive of doing science is mixed with other *non*scientific motives – no wonder complexities arise. Let us, then, confine ourselves to science being pursued *solely* for the sake of science. Even under such a restriction, I contend, the problem of group rationality has no obvious solution (with or without the utilitarian assumption). (*b*) The success of the ends that these nonscientific motives are pursuing is also essentially dependent on achieving truth.

39 And well he might: "Why should it be *uniquely* rational to pursue one's own self-interest to the exclusion of everything else? It may not, of course, be at all absurd to claim that maximization of self-interest is not irrational, at least not necessarily so, but to argue that anything other than maximizing self-interest must be irrational seems altogether extraordinary." Sen, *On Ethics and Economics*, 15; see also Sen and Williams, *Utilitarianism and Beyond*, 4–5. My defense is (*a*) that at least in the context of the problem of group rationality, it is not irrational, as is granted, to assume maximization of self-interest, and (*b*) not that heeding the interests of other scientists is irrational, but that its rationality, in light of (*a*), needs justification, and cannot simply be assumed. I conjecture that providing that justification will unveil some pretty startling assumptions, assumptions that measure the depth of the problem of group rationality.

Such a counter would be deeply unsatisfying. Here's why. Let us say, with Sen, that the medley of motives that scientists have – some of them are *self*-directed, some are *other*-directed – includes motives not only to advance their own respective fields but also to help other fields develop. But the scientists would be unwilling to sacrifice entirely their own respective fields for the welfare of other fields; if they were willing to so sacrifice – or if they were quite unwilling to do so at all – there could be a solution at hand in the first case, none at all in the second. The solution to the Sen-problem of group rationality demands that there be a *principled* way in which scientists can collectively agree to structure their society so that not only each individual scientist benefits from that cooperation,[40] but also the resultant *science* is better off than under any other alternative matrix or structure.[41] Thus, it is not merely a matter of respecting the rights of fellow scientists, or their interests, or what would satisfy them, but a problem of what distribution of scientific labor would make *science* the beneficiary.[42]

S_2 (*The Sen-Problem of Group Rationality*): How should *n* scientists structure their activities so that the scientific welfare is maximized (where what is jointly produced is shared without remainder for consumption)?

[40] Condorcet, at least, contends – and there is little in Smith to contradict it – that each individual economic agent has not only desires and goals, but also a theory of how things ought to be; see Rothschild, *Economic Sentiments*, 157. This significant assumption must be interwoven into an adequate theory of group rationality. One can see scientists bringing to the table not only the goals they wish to pursue, the problems they want to solve, the theories they wish to extend, the experiments they mean to conduct, the observations they want to make, the data they plan to collect, but also theories of how their group should be structured. That said, one must still find – a nice task – a meta-theory that will carefully chip away at their theories of, or interests in, how the group should be structured through consensus. Or else the scientists will be hopelessly deadlocked.

[41] To be distinguished from what would make scientists happy, increase their utility, satisfy their desires, and so on.

[42] In arguing against George Stigler's "Economics or Ethics?" Sen has claimed that in fact little empirical testing of the hypothesis – namely, that when self-interest and ethical values conflict, the former will always win – has been done. He says, "the success of a free market does not tell us anything at all about what *motivation* lies behind the action of economic agents in such an economy." *On Ethics and Economics*, 18. Not only that, Sen offers the example of Japan, where in fact departure from self-interest in the direction of duty, loyalty, and goodwill has led to industrial success.

It is not my place to argue with Sen and Stigler on economic issues. But, as a moral philosopher, I might ask, "Would duty, loyalty, and goodwill have prevailed in the *absence* of industrial success in Japan?" Suppose *not*: One might then argue that the

The solution to the problem S_1 – it is worth noting – will not generally be equal (except fortuitously) to the solution to the problem S_2. For what maximizes the scientific welfare *as a whole* need not maximize the respective self-centered scientific welfare of each scientist in the group (defined solely in terms of *his science*); thus, a scientist may wish to advance the cause of quantum field theory to an extent that would not allow other significant fields to be developed in the interest of science as a whole. And, of course, vice versa. Yet this simple, significant inequality is rarely, if ever, noticed – assuming that the two problems are even distinguished. If that argument is sound, then it would immediately follow that the solution to neither S_1 nor S_2 would entail a solution to **UGR** (which involves the happiness of a scientist in the pursuit of his science).

Let us say that there are n scientists in the group and m ways, S_1, S_2, $S_3, \ldots, S_m$, in which their cooperation can be structured,[43] resulting, respectively, in the promise of P_1, P_2, $P_3, \ldots, P_o$ scientific product.[44]

motives of duty, loyalty, and goodwill are ultimately underwritten by self-interest. Then again, suppose *yes*: It would follow that duty, loyalty, and goodwill do not function essentially in industrial success.

My ultimate interest here, however, is neither in economics nor in ethics, but in science. If being interested in the science that matters to fellow scientists is what makes for group cohesiveness and successful science, that needs to be *shown* – in particular, it must be shown what principles of rationality would lead a scientist to pay equal, or at least some, attention to the welfare of the science his fellow scientists are interested in rather than looking out exclusively for his own scientific interest. Alternatively, we have to show how exclusive attention to his own scientific interest will lead each scientist to give his allegiance to the group under a certain institution or matrix, thus leading to a cohesive group and successful science. See Chapter 9 of this volume, section II, which delves further into this problem. This issue is also connected to what I have called the problem of fragmentation; see Chapter 4, 81–2, and the references in footnote 19.

[43] How these ways are to be specified is a separate, highly interesting question. Freedom gets defined not only in terms of alternatives that are available (or lost), but also in terms of the *goodness* of these alternatives; Sen, *The Standard of Living*, 36 and footnote 17. In a similar fashion, one might say that the freedom of the scientists is defined not only in terms of the alternative institutions or matrices available to them under which they will agree to cooperate, but also in terms of how *rational* these institutions or matrices are.

[44] Herein lies another nice difference between science and welfare economics. If one wants more housing, medical care, transport, roads, and such, one can reasonably well and clearly *specify* what is needed, and while it may be a difficult task to determine the means to those ends, the ends themselves are not ambiguous. In science, by contrast, beyond saying something vague and general like "We need an answer to the

Assuming an *independent* way of evaluating these products or states – without resorting, that is, to what the scientists want, or what they seek satisfaction in (the *heterogenous product*), and so on – and assuming that S_O is the optimum institutional structure yielding optimum science, P_O (defined in terms of verisimilitude), then the Sen-problem of group rationality is to show *how* scientists would be led to agree – *by what principles of rationality* – to harness their efforts under the structure S_O to produce P_O. Confining this task by making each scientist a maximizer of his or her scientific goal will lead to a cul de sac – for the well-worn reason that we would be trying simultaneously to maximize *n* variables (scientific goals). We must make each scientist in the group interested in the welfare of fellow scientists as well. Consequently, neither *self-centered scientific welfare*, nor *self-scientific welfare goal*, nor *self-scientific-goal choice* can be retained without radical revision. Thus far, this is, I think, an unabashedly Senian conclusion.

Yet without a proper rationale, arbitrarily introducing an interest in one's fellow scientists, into the motivational repertoire of a scientist in determining what he should do would be utterly ad hoc. Consequently, our task would have to be to show *how* and *why* a scientist might be interested not only in his own welfare, but also in the welfare of fellow scientists; and the extent or degree of that interest in the goals of other scientists. Finally, given the various ways in which the scientific matrices or institutions, S_1, S_2, $S_3, \ldots, S_m$, might be created to accommodate various interests, to which matrix or institution should a scientist give his allegiance and, more importantly, why?

An expansive variant on the Sen-problem of group rationality S_1 is:

S'_1 (*The Sen-Problem of Group Rationality*): How should *n* scientists structure their activities, given that there are *m* ways, S_1, S_2, $S_3, \ldots, S_m$, in which their cooperation can be structured, resulting, respectively, in the promise of P_1, P_2, $P_3, \ldots, P_o$ scientific product, so that the result of their cooperation leads each scientist to maximize his or her own self-centered scientific welfare, the only principle of rationality to be invoked (and where what is jointly produced is shared without remainder for consumption)?

biological question 'What is sex for?'" – for example – we have no clue about what the correct answer will even look like. There is no specifying the results of science in advance.

It is obvious that the scientists are not going to cooperate under the structure or matrix S_i if that means that one of them maximizes his scientific goal while the rest relinquish maximizing theirs in favor of his. So a structure, S_O, will have to be negotiated in deviating from which no one can get the cooperation of others. What makes S_O the optimum solution? How is optimum here to be defined? What principles of rationality will have to be invoked to get each scientist to work under the aegis of S_O?[45] Notice: Looked at purely from the point of view of science, this may not be the optimum structure.

An expansive variant on the Sen-problem of group rationality $\boldsymbol{S_2}$ is:

$\boldsymbol{S'_2}$ (*The Sen-Problem of Group Rationality*): How should *n* scientists structure their activities, given that each scientist is involved in directly attaching importance to the welfare of other scientists in the group, and that there are *m* ways, $S^*_1, S^*_2, S^*_3, \ldots, S^*_m$, in which their cooperation can be structured, resulting, respectively, in the promise of $P^*_1, P^*_2, P^*_3, \ldots, P^*_O$ scientific product (and where what is jointly produced is shared without remainder for consumption)?

"It would be extraordinary," Sen observes, "if self-interest were not to play quite a major part in a great many decisions, and indeed normal economic transactions would break down if self-interest played no substantial part at all in our choices."[46] Not only economic transactions, one wants to add, but scientific transactions as well. No scientist will – should? – relinquish all his scientific interests in favor of fellow scientists' interests (even to maximize the benefit for science per se?).[47]

[45] This question highlights the importance of individual rationality.

[46] Sen, *On Ethics and Economics*, 19.

[47] Smith and Condorcet thought that agents have a multiplicity of aims and goals; this multiplicity will lead to conflict, and there is a plea for a "civilized conflict." There are two ways to look at the solution to this problem: *either* as one imposing a pattern or an end state, *or* as one aiming for a state – whatever it turns out to be – that is a fallout upon the mutual agreement of the parties. Let us call the first *the end-state solution*, the second *the agreement solution*. It appears that Smith and Condorcet would have favored the agreement solution. The former type of solution, they would have said, hints at too much interference and restriction of liberties; it might be a cause of vexation. See Rothschild, *Economic Sentiments*, 197–203.

Ultimately, the solution we are seeking should be an end-state solution, although based upon mutual agreement. An analogy will make for a tidier, if not more convincing, presentation of the idea. Let us suppose that we are concerned with a just state; Rawls's difference principle is what is agreed upon in the original state. Once out of that original state, moral agents will know what they should be working *toward*, if they are not already in a just state. Similarly, there is a state in which the task of

Consequently, the task is to show – on the basis of what principles of rationality? – why, of the several structures, S^*_O will be settled upon, a structure in which each scientist can feel that *his* particular scientific interest has been generally given a fair hearing while the structure, has simultaneously heeded the scientific interests of *others.* Once again, S^*_O's optimality will have to be defined.

IV. The Problem Defined

I have tried to show that Sen offers us deep, marvelous, and myriad ways by which to analyze the problem of group rationality. It is time, then, to turn to the solutions, to examine some of the extant theories of group rationality. We will let others decide whether evolutionary and game-theoretic approaches to that problem triumph over the standard philosophical approach; and if they do so decide, we will let them set about showing us just how and why. Until then, we shall proceed on the assumption that the problem of group rationality is unique and that the use of the standard philosophical approach will yield a rich harvest.

What, then, is that unique problem of group rationality? It is this:

> Engaged in a joint enterprise to pursue a set of shared (partially overlapping?) goals, what ought to be the structure of a society of scientists – the structure defined in terms of method – at a given time, or over an interval of time, in virtue of which that structure would best enable that society of scientists to reach its goals better than it would under any other structure?

The solutions that follow – namely, the skeptical view, two versions of the subjectivist view, the objectivist view, and Putnam's view – will be examined in light of this formulation of the problem of group rationality. Even so, I shall tread cautiously. Paying due deference to the history, character, and distinctiveness of each solution, I shall reconstruct the

science can be performed most effectively; to be sure, it is a state that has to be argued about, discussed, evaluated. Once an agreement has been reached (*how* it is reached is quintessentially a different, significant issue), the scientists will know what they need to do in order to reach that state. In the absence of that knowledge – or agreement, at the least – scientists will not have a yardstick by which to measure the rationality of their group. *Whatever* agreement they chance upon will be right, and then right will lose its normative force. Here I depart from Rawls's constructivism; see *A Theory of Justice*, especially 40.

problem of group rationality from the vantage point of the particular solution in question, and only then examine both that problem and the solution in light of our own foregoing formulation of the problem of group rationality. Our task, then, is dual: first, to reconstruct some of the classical theories as solutions to the problem of group rationality (in itself a novel, interesting task); and second, to discover which of these theories, if any, adequately solves that problem.

4

The Skeptical View

The skeptic's view is unreasonable.[1] This we know from the start. Why, then, does one examine it? The dogmatist's view represents the limit at which reason has not seriously commenced. The skeptic's view represents the limit beyond which our reasoning cannot go. Between these two views there is room enough for philosophy. The function of the skeptic's view is to pull us as far away from the limit of dogma toward itself; witness Descartes' First Meditation. So it is in our interest to present as powerfully as we can the case of the skeptic, to present his view as a challenge to every step and substep of the argument – in this case, an argument for a traditional theory of group rationality wherein reason is cast in the king's role. The case thus presented, we should know what challenges are to be met, and if we have met them; what problems are to be solved, and if we have solved them. One measure

[1] Throughout this chapter, I have relied on the works of Paul Feyerabend, and yet I have not mentioned him by name in the text. The reasons are two. First, he has little to say about the problem of group rationality; consequently, what I have done is to reconstruct his views to serve my purposes. Second, Feyerabend has never found anyone – friend or critic – really able to understand what he is about. An excellent reason, I thought, to believe that he would be equally dissatisfied with my presentation of his views. Not wishing to be embroiled in a mere exegetical debate, I have left his name out. I leave the reader to decide whether I have been fair and, far more importantly, to decide whether the skeptic's view charted here is philosophically as useful as it might have been.

There are, of course, other methodologies that appear in more traditional guises, but that are in fact as skeptical as the view treated here. For an analysis of some of these other methodologies, see Husain Sarkar, "Anti-Realism against Methodology."

of our success, or near-success, must consist in showing how the view we offer is judicious, balanced, and plausible, while the skeptic's view is not.

I begin by outlining, in section I, the skeptic's notion of the Democratic Councils. Here I portray the essential tension between a scientist's scientific interests and his or her other interests, a tension in which, as the skeptic sees it, the latter interests should not yield to the former. Section II depicts why the skeptic thinks a scientist should not enter into any covenant concerning aims with any other scientist, thus proliferating the aims of science. Section III focuses on the skeptic's view of why scientists should produce an endless multiplicity of theories, metaphysics, and methods. The next two sections ask if the skeptic suggests the following of any method, or of no method, as a way of structuring the scientific society. They do so under the respective headings of positive and negative arguments. Then the last section VI raises the means-end issue, namely, how should the society of scientists, as the skeptic envisages it, be brought about? In short, it outlines how the skeptic might solve the Williams problem.

I. The Democratic Councils

The skeptic's idea of the Democratic Councils is of first importance.[2] Here is how and why. The primary question in political philosophy is not "How extensive should be the right of the state?" but rather "Should there be a state at all?" Some political anarchists want to argue for a stateless society as being the only just society; others argue for a minimal state as being compatible with the rights of its individual citizens, and hence just only so far. Correspondingly, one might say that the primary question in the field of group rationality is not "What should be the basic structure of the scientific society in virtue of which it is rational?" but, on the contrary, "Should there be any structure at all?"

[2] I stand cautioned: Danish-style consensus conferences are not identical to parliaments to which representatives are elected. Since Feyerabend doesn't draw any such distinction, let alone defend one particular version of a Democratic Council, I have let the term stand in representing the skeptic's views. Perhaps, based on Feyerabend's anarchistic views, these are also up for discussion by the electorate?

The skeptic wants to argue for a structureless society as being the only one that ensures a rational society: a society that is conducive to the fulfillment of the variegated aims of science. To impose a structure on the society that hampers the free movement and exchange of any thoughts, ideas, experiments, criticisms, and queries from a multiplicity of perspectives is not rational at all, because such a society will be marked by a failure to reach its multifarious aims. If the primary subject of rationality is the basic structure of the scientific society, and if the basic structure is defined in terms of method, then only that scientific society that has no method at all is rational. No structure, no method. Or, if it is to have any structure at all, then it is defined by the principle "Anything goes." Any structure, any method.

First, from the point of view of the skeptic, the problem of group rationality is a genuine one. Second, his solution to the problem is the principle that anything and everything should be allowed or be made permissible. Finally, he provides us arguments that show that only such a society would lead to whatever aims of science are postulated. Our task is to examine those claims.

The skeptic argues that in order to determine which theories or traditions to pursue and which to support with public money, the society should set up Democratic Councils. It is the people, through their elected representatives, who should decide and determine which scientific programs, theories, and traditions are worthy of pursuit and also in the public interest, and which theories and traditions should be abandoned or rejected, at least in part as being against that public interest. Consequently, part of the solution to the problem of group rationality is the introduction of the Democratic Councils. For the skeptic, the rights of the individual citizen, as well as his flourishing, are of far greater importance than such things as beauty and truth. But is the lay individual competent to decide such matters? The skeptic would side with Étienne Geoffroy Saint-Hilaire who, in nineteenth-century France, often appealed "to the public rather than to the academicians as the final arbiter on scientific matters." He gave this as his reason: "I have faith in the solidity of popular judgments. The masses are endowed with an instinctive sense, which renders them shrewd and makes them very skillful at seizing the synthetic point of questions."

Ordinary scientists are lost in minutiae and cannot form an opinion based on a larger view of things.[3]

(Parenthetically, John Stuart Mill would have found this to be a remarkable idea. Mill himself was unsure of the reasonableness of the outcome of the democratic procedure: He was not confident that the mass of biased and uneducated opinions would produce the desirable outcome. Thus, in his *Thoughts on Parliamentary Reform,* he advocated the "Hare Plan." In part, the plan was intended to ensure that a dispersed group of individuals with a common conviction, but whose number equaled the size of a parliamentary constituency, could send a representative to the Parliament to air their views. In his *Considerations on Representative Government,* he advocated that educated citizens should have plural votes and that professional civil servants should have a larger say in policy making and administration. This from the author of *On Liberty.* Is it to be doubted that he would not have approved of the skeptic's idea of Democratic Councils as a way of deciding on technical matters in science? Or, at the very least, would he not have advocated giving multiple votes to scientists in matters of science?)

Let us expand this claim. There is the scientific society, and here are the theories or traditions among which some have to be selected for pursuit. Some controls and constraints are required to select the theories or traditions. *Who* should impose the controls? And *what* controls should be imposed? The answer to the first question, according to the skeptic, is "*fundamental debates between traditions are debates between laymen which can and should be settled by no higher authority than the authority of the laymen, i.e., democratic councils.*"[4] In short, laymen should impose controls, and whatever controls are imposed by the laymen can and should be deemed proper.

[3] Yet George Sand feared to tread where Geoffroy would have the populace rush in. Geoffroy frequently sent Sand copies of his books and articles and begged her to circulate his ideas and theories through her writings. The novelist refused: "I would surely write a laughable work if I tried to explain how all that arranges itself in my head; thus I will not undertake it." See Tobby A. Appel, *The Cuvier-Geoffroy Debate: French Biology in the Decades before Darwin,* 187 and 189. See also the poignant passage quoted from *Education for Exclusion or Participation?* by the Iranian scholar Majid Rahnema, in Feyerabend, *Farewell to Reason,* 298.

[4] Paul Feyerabend, *Problems of Empiricism,* 32.

The skeptic contrasts his Democratic Councils with the elitist views of others. According to one elitist view, each scientist in his own special discipline is the best judge; no outsider, no ordinary person can do better. Consider the fields of radio astronomy, genetic engineering, and elementary particle physics; an ordinary citizen cannot even superficially understand the problems, let alone distinguish central problems from peripheral ones, much less understand competing theories in the field and evaluate their potential for further research. The knowledge of the field is born through dwelling in the particulars, as Polanyi puts it, and cannot be articulated and hence cannot be transmitted to ordinary women and men. This tacit knowledge grows and accumulates among practitioners of a common field and enables them to make decisions that no outsider can make.

An alternative elitist view, by contrast, stakes its claims on explicit knowledge; methodological standards, it avers, must be made explicit, and decisions concerning which theories to accept or reject, for example, must be based on these overt standards. This elitist view regards as important only certain aspects and fields of science – such as Newtonian mechanics, the special and general theories of relativity, the transition from the Ptolemaic to the Copernican theory, and Darwin's theory of natural selection and evolution – while other aspects and fields – such as the fields of empirical sociology, psychoanalysis, astrology, and parapsychology – are to be regarded as at best lying on the fringes of science. Reason, through the study of such disparate disciplines, will enable us to discover what is science and what is not, what are and are not the proper standards for evaluation of scientific traditions and theories, and so on; hence the significance of reason and the tradition of rationalism. On this view, then, the standards can, and should, be made explicit, while they are, and can only be, implicit on the other elitist view. Furthermore, these standards can in principle be separated from the practice of science. Dwelling in particulars may be of heuristic value, but it is not essential to understanding the methodological standards; thus once these standards are articulated, ordinary citizens can understand and apply them. Lastly, these standards are general in scope: They apply just as well to cosmological theories as they do to theories in cytology.

The skeptic disagrees. "A free society must not be left at the mercy of the institutions it contains; it must be able to supervise and control them."[5] Once the church dominated society with disastrous consequences, and this happily led in some free societies to the separation of the state and the church. In our own time, the institution of science is dominating every aspect of our lives, often unbeknownst to ourselves, and it is time that a free society also spoke of a separation of the state and science. Elitism in science, says the skeptic, is as unnecessary and pernicious as elitism in religion. The elitism of the scientists, on the first view, threatens the autonomy of science and the lay scientists, just as the church did at one time.[6] The second view, on the other hand, is based on an inadequate foundation. It assumes that standards, articulated or otherwise, cannot change. But standards can, and do, change, and they play a significant role in the growth of knowledge. It assumes that standards define our rationalism. But science and rationalism should not be regarded as standards by which to judge other traditions, for they are themselves traditions on a par with the ones they would judge. "Rationalism is not a boundary condition for traditions; it is itself a tradition, and not always a successful one."[7]

There are two questions to be asked and answered. First, is the Democratic Council competent to decide scientific questions? Second, if not, is there a danger of confusing two questions? Let me take the second query first. We must distinguish the question "What is a just society?" from the question "What is a rational scientific society?" The answers to the two questions do not coincide. A just society is not necessarily the happiest society, for it is not the aim of a just society to maximize happiness (although it may do so quite accidentally). Similarly, a just society is not necessarily a rational scientific society, for it is not the aim of a just society to fulfill the aims of science. It is entirely possible, perhaps even likely, that a just society that violates no one's rights and lets individuals pursue their own aims, interests, and inclinations in a manner they deem appropriate is not a rational society – in other words, not a society that would successfully lead to fulfilling the aims of science.

[5] Ibid., 25.

[6] "Of course, scientists will not play any predominant role in the society I envisage. They will be more than balanced by magicians, or priests, or astrologers." Paul Feyerabend, "How To Defend Society against Science," in *Knowledge, Science, and Relativism*, 190.

[7] Feyerabend, *Problems of Empiricism*, 27.

Perhaps it might be useful here to recall an earlier formulation of the problem of group rationality.[8]

UGR (*The Utilitarian Problem of Group Rationality*): How should *n* scientists structure their activities, given that each scientist is involved in directly attaching importance to his or her own scientific welfare, and the happiness he or she derives from that welfare, in doing, testing, and producing science (and where what is jointly produced is shared without remainder for consumption)?

Notice that this formulation is a bit more circumspect than the skeptic's. The happiness of the scientist is closely tied to what drives him, namely, the welfare of his science; it is not just happiness generally speaking. Given that constraint, we can ask, "What would make a scientist happy?" Presumably, a successful theory. Now, while it *may* be plausible for each moral agent to be the ultimate judge of what makes him happy, no scientist can – or should – deem himself to be the ultimate arbitrator of what constitutes a successful scientific theory. If he is to be happy in his science, his *science* must be successful.[9] But whether it is successful or not is not for him to say; it is the verdict of an independent reality. Consequently, there is one remarkable parallel between scientists and moral agents: In the latter case, good deeds are not always rewarded; in the former, hard scientific work does not always meet with success. Whatever may be the case in morality, in science there is an independent reality to which a scientist's theory must answer.

The skeptic might reply that it is far more important to allow humans to flourish and have liberty than to have a society in which science prospers – with reality serving as an independent touchstone – at the expense of things we usually value. So, of the two competing aims – science, on the one hand, and human rights and flourishing, on the other – the skeptic accords priority to the second aim.[10] But then it is quite clear that the skeptic's view is not a solution to the problem of group rationality, since it is conceded that in a society as he envisages

[8] See Chapter 3 of this volume.

[9] "But so far as truthfulness is concerned, to make a lot of the fact that scientists' individual motives are more worldly than the Platonic myth suggested is significantly to miss the point. Their goal is fame, above all fame and prestige in the scientific community itself, and that will come from the recognition that they have done good science." Bernard Williams, *Truth and Truthfulness*, 142.

[10] See also Feyerabend, "How to Defend Society Against Science," in *Knowledge, Science, and Relativism*, 190–1.

it, while humans may flourish and rights may be left untrammeled, science may well suffer, in the sense that it becomes too chaotic an enterprise to achieve its aims. Note: The skeptic himself offers his view as an excellent medicine for epistemology, not as a solution to a problem in political philosophy.

The skeptic's view enables us to sharpen our understanding of the problem of group rationality,[11] and here is how it does that. Assume a free, anarchist society. Assume that a group of individuals is interested in science and knowledge as others in the society are interested in the arts, drama, poetry, music, and such. Now, if the society is free, there will be freedom of association.[12] To associate with the group of scientists, one needs to qualify. The qualified members of the group can then ask themselves the question "How shall we organize in order to acquire as much knowledge as we can?" To that question, the skeptic has no answer. Or, if he has one, it is at least problematic.

The skeptic can no longer argue against this society as not conducive to human flourishing, for the scientists who form the association have themselves chosen to flourish as scientists; the skeptic can no longer argue against this society as transgressing human rights and liberty, since the forming of the group was *ex hypothesi* a free association. No one is dragged into the society and ordered to follow its rules and injunctions; no one is left out of the association, if he agrees to subscribe to the norms of the association. But such a society will *still*

[11] The problem was formulated earlier at the end of Chapter 3.

[12] The scientific society, if it is to be a free association in a free society, should not be hampered in its form and function by the society at large, or else the society is not free by the standards of the skeptic or the anarchist. If anarchy is a Social Utopia – from a skeptic's view – then will epistemological anarchy mirror Scientific Utopia? Arguably, the skeptic thinks so.

Now, even such a scientific society will have constraints, especially economic constraints. So long as such a society is not dependent for its economic resources on the society at large (an utterly unrealistic assumption), it will not be beholden to it and, hence, will not need to rely on the decisions of the Democratic Councils of that society. Even under such an unrealistic assumption, the problem of group rationality does not go away. Of course, the economic constraints are not the only side constraints; psychological, sociological, and other such factors will constrain the form and structure of the scientific society, too, constraints over which the scientists may have far less control. Again, even when these side constraints are satisfied, the scientists will still be confronted with the problem of how they should structure their society.

face the problem of group rationality. Is it the case that the skeptic's solution is an adequate solution for *this* society? Would its members espouse the solution that their society should permit the rule "Anything goes" as its only principle? This way of viewing the matter weeds out the irrelevancies in the skeptic's position and presents a sharpened version of the problem.

II. No Covenant, a Tale

Here is the second challenge of the skeptic. There should be *no* covenant between the scientists in a scientific society. For the scientists should be open to a proliferation of aims, methods, and theories. The skeptic:

> It is clear, then, that the idea of a fixed method, or of a fixed theory of rationality, rests on too naive a view of man and his social surroundings. To those who look at the rich material provided by history, and who are not intent on impoverishing it in order to please their lower instincts, their craving for intellectual security in the form of clarity, precision, 'objectivity', 'truth', it will become clear that there is only one principle that can be defended under *all* circumstances and in *all* stages of human development. It is the principle: *anything goes.*[13]

[13] Feyerabend, *Against Method*, 27–8.

Three notes. First, a note about *Against Method*: Three versions under that title have been published: in 1975, 1978, and 1988, respectively, not to mention the third German edition of *Against Method* published in 1986, "which," says Feyerabend, "differs from the English, the French, the Japanese and the Portugese editions" (*Farewell to Reason*, 280). There is surely some truth in Ian Hacking's remark that these are three different books with the same title rather than one book undergoing successive revisions; see his "Paul Feyerabend, Humanist." Since what I am doing is not exegesis, I have not attempted to explain how Feyerabend's views evolved in these texts. Where I refer to an edition other than the 1975 one, I have specified it.

Second, since I do refer to *Science in a Free Society*, I feel obliged to note that we now know that Feyerabend, in the last ten years of his life, was sufficiently unhappy with *Science in a Free Society* to refuse to have it reprinted; see Paul Feyerabend, *Conquest of Abundance*, xi. What is curious is that although *Farewell to Reason*, for which Feyerabend was awarded the Premio Fregene in Italy in 1990, has a deep and vigorous affinity to *Science in a Free Society*, Feyerabend, so far as I can tell, expressed no diffidence about that work.

Third, for Feyerabend's own account of the genesis of *Against Method*, and his response to some of the reactions against that book, see Chapter 12 of *Killing Time: The Autobiography of Paul Feyerabend*.

The skeptic again:

> Knowledge so conceived is not a series of self-consistent theories that converges towards an ideal view; it is not a gradual approach to the truth. It is rather an ever increasing *ocean of mutually incompatible (and perhaps even incommensurable) alternatives,* each single theory, each fairy tale, each myth is part of the collection forcing the others into greater articulation and all of them contributing, via this process of competition, to the development of our consciousness. Nothing is ever settled, no view can ever be omitted from a comprehensive account.... The task of the scientist is no longer 'to search for the truth,' or 'to praise god', or 'to systematize observations', or 'to improve predictions'. These are but side effects of an activity to which his attention is now mainly directed and which is '*to make the weaker case the stronger*' as the sophists said, *and thereby to sustain the motion of the whole.*[14]

These passages clearly indicate that in a scientific society there ought to be no fixity in aims, no fixity in theories, and no fixity in methods. I will return to theories and methods later. Here I wish to concentrate on the collective aims of the scientists. From the point of view of the skeptic, there should be no common aim toward which the scientific society as a whole should be moving. Neither truth, nor predictability, nor empirical adequacy alone should be the aim. The aim should simply be to fashion a plethora of mutually incompatible theories; far from abandoning weaker theories, the aim should be to try and invigorate them; that should be an end in itself.

Let me weave a tale.[15] The tale is designed to render perspicuous the skeptic's aim of science or knowledge, his principle of extreme proliferation, and his view of structuring a rational scientific society. Imagine: There is a convocation of *all* the scientists. Those working now, in the twentieth century – such as the evolutionists, quantum physicists, molecular biologists, geochemists, weed scientists, ecologists, anthropologists, cosmologists, and plant pathologists – *and* scientists of the past, such as physicists, chemists, alchemists, astrologers, astronomers, geographers, barber surgeons, apothecaries, physicians, botanists, herbalists, and geologists. These are resurrected for the

[14] Feyerabend, *Against Method,* 30.

[15] The tale is, in some small measure, inspired by the one told by Richard Rorty, Jerome Schneewind, and Quentin Skinner in the opening pages of the introduction to their edited volume, *Philosophy in History: Essays on the Historiography of Philosophy.*

convocation. The aim of staging the convocation is to produce a massive collection of volumes called *The Books of the Sciences.*

Here is how. Each scientist parleys with each of the others, reads everything the others have written, and the compliment is returned. They then gather – there being neither temporal nor geographical restrictions – for conversation. Paracelsus will encounter not only his archenemy Galen, but also the blind Waichi Sugiyama, the distinguished 17th-century Japanese acupuncturist, and Galileo's radical father, whose expertise lay in the field of music theory. Euclid will pair not only with Archimedes and Kurt Gödel, but also with John Dee and ibn-Shatir. Thales will hook up with Qusta ibn Luqa as well as William Gilbert and Sacrobosco; Newton not only with Planck and Riemann, but also with al-Faraghni and Yang Chi-chou; Charles Darwin with al-Razi and Chandrashekhar; James Clerk Maxwell with Hunayn ibn Ishaq[16] and Ramanujan. Copernicus with John Maynard Keynes. Eudoxus with al-Khawarizmi and John Ray; Georges Cuvier not only with Aristotle and Vesalius but Freud and Condillac as well; Emile Durkheim not only with Veblen but also with Nagarjuna and Gauss. And so on and on and on.

Their readings and conversations will produce works of unparalleled interest. For each round of meetings will produce a new book, a new paper, a new discovery, a new thought, a new experiment, a new problem, which are then considered in the next round, and so on. These now belong to *The Books of the Sciences*, and the method of permutation and combination assures us that the process of adding to *The Books of the Sciences* will never cease.

The skeptic is not suggesting that we will have moved closer to the truth, or that we shall have enhanced our ability to make predictions, or any such thing; he only claims that the society of scientists will have fulfilled the skeptic's aim: There will be an expansion of the sum of mutually incompatible alternatives. Any restriction on aims, on the skeptic's view, will have a deleterious effect on knowledge. For instance, Stephen Jay Gould may not have the same aim of science as did Marcello Malpighi or Thomas Huxley, even in the same discipline, let alone that Maestlin should have the same aim of science as Gould when he is conversing with Gould on the phases of Venus. Avicenna's

[16] The son of a Christian druggist of the ninth century.

view of the science of logic may differ, and it did, from Gottlob Frege's. Hence, if some restriction were antecedently placed on who could converse with whom, then interesting ideas of scientists with widely different aims would not be allowed to clash and develop, and *The Books of the Sciences* would thereby be scantier and impoverished. *This is the reason, from the point of view of the skeptic, why there should be no covenant of aims in the Democratic Councils.*

Now, the tale was designed to illustrate the skeptic's *aim* of science – namely, vast, incompatible theoretical alternatives ought to constitute the aim of the scientific society. Consequently, the problem of group rationality must be stated not in terms of the Peircean ideal state of knowledge or the realist aim of science, or any other such notion, but rather in terms of the skeptic's aim of producing *The Books of the Sciences.* So the skeptic would redefine or reformulate the problem of group rationality in this way:

> *The Skeptic's Problem of Group Rationality*: Engaged in a joint enterprise with a multiplicity of aims, in part to produce an endless string of theories that are mutually incompatible, how ought the scientists to structure their society so that they would reach their goals better than they would under any other structure?

Indeed, given the plethora of aims, it would seem that perhaps the only way to satisfactorily reach these aims is by way of a plethora of theories and methods; in short, by way of the skeptic's anarchism. This is, in fact, as we shall see, the skeptic's solution to the problem of group rationality.

Before we turn to his solution, let us briefly note some similarities and differences between his formulation of the problem and ours.[17] (*a*) Both we and the skeptic define structure in terms of method; (*b*) in our formulation of the problem, it is presupposed that scientists pursue a set of shared – or at least partially overlapping – goals; by contrast, the skeptic avows that there should be no covenant between scientists on aims (or on anything else); and finally, (*c*) our formulation of the problem presupposes that the scientists are engaged in a joint enterprise explicable in terms of shared goals; it is a nice problem for the skeptic's view, given his formulation of the problem, what

[17] For my formulation, see Chapter 3 of this volume.

makes the scientific enterprise *joint*. As we shall see, the latter is not an innocuous problem.

Let us, then, examine the adequacy of the aim of science proposed by the skeptic and in light of which he formulates the problem of group rationality. First, there is the simple logical problem. Mutually incompatible theories are logically inconsistent. During every period in the history of science, the sum total of extant theories were mutually inconsistent or, at the very least, contraries. Thales thought everything was water; Anaximander thought everything was air. Newton thought light was corpuscular in nature; Young and Fresnel thought it consisted of waves, and so on. Consequently, far from postulating an unreachable aim, such as the aim of truth, the skeptic proposes an aim that we have repeatedly reached. Now, why should one prefer one inconsistent or mutually incompatible set of theories over another? Does not everything follow from one inconsistent set?[18] In short, the skeptic's version of the problem of group rationality has already been solved in practice, *given* the skeptic's aim. The skeptic's failure lies in providing no rationale for the aim he postulates.

Second, the skeptic's view runs into a problem that I shall call, given its significance, *the problem of fragmentation*.[19] The skeptic cannot object to any scientist pursuing any theory with a view to fulfilling any aim. The skeptic says so in so many words. Hence, the aim of producing mutually incompatible theories is simply an unintended consequence of the group following a plurality of aims. The skeptic evidently fails to account for why there is, can be, or ought to be a coordinated group activity. In virtue of *what* single or several aims are the scientists said to be forming a *group*? Why should we not refer to the society of scientists as a splinter group whose subgroups – or individual scientists – have nothing in common? For instance, what do scientists who belong to the subgroup pursuing truth have to gain from scientists pursuing aesthetically pleasing theories? The subgroups are fragmented, given their incommensurable aims; they might just as well be regarded as groups unto themselves, each subgroup, given its aims, facing a distinct

18 *Pace* logicians who claim that contradictions can be true. Also compare Chapter 3, footnote 1, this volume. For a different set of criticisms of Feyerabend on consistency, see John Watkins, "Feyerabend among Popperians 1948–1978," especially 52–4.

19 See also 106, 111, 246–7.

problem of group rationality. Going further still, each scientist might just as well be regarded as a group unto himself.

Third, and centrally, we ask, why an endless plethora of aims? The skeptic arrives at his skeptical position because of a common pattern of his demands for justification, and these demands are formidable. To justify a proposed aim is to demonstrate that (*a*) the aim can be made philosophically clear and coherent, (*b*) the aim can be reached in principle, and (*c*) the proposed solution to the problem of group rationality is an assured or guaranteed way of reaching that aim. To fail on any one of these three conditions, on the skeptic's view, is simply to fail to justify. Consequently, failure to justify any particular aim is evidence in the skeptic's eyes only that that aim is on a par with any other similarly unjustifiable aim.

Let us see how the skeptic's view fares on his own view of justification. The skeptic, I shall argue, is hoisted by his own petard. First, there is no clear delineation of the aim of mutually incompatible theories. What is the significance of this aim, particularly since reaching it entails reaching necessarily false statements? What purpose or end does this serve as means? What is the ultimate purpose or end? Why should it be regarded – like truth or beauty – as intrinsically valuable or important in its own right?

The ultimate end is the proliferation of final aims; the pursuit of these final aims unintentionally (or deliberately) produces mutually incompatible theories. The society of scientists is then viewed as a conglomeration of various subgroups, each defined by its own distinct aim. Some subgroups pursue the aim of truth, some pursue simplicity, some aesthetically pleasing theories, and still others pragmatic ones. *None* of these aims is problem-free and clearly defined. Then, instead of having one problem of defining – say, the aim of truth – for the scientific group, the skeptic has several problems, none of which are resolved, or justified, to the degree the skeptic insists on. Thus, the skeptic's view fails on its own first condition of justification, namely, (*a*).

Next, can the skeptic show that the aim is reachable? Since we have mutually incompatible theories all the time – the skeptic's aim being trivially satisfied – why do scientists continue to pursue their activities? On the other hand, if the skeptic's aim is nontrivial, he offers no argument to show that it is reachable. Hence, the skeptic's view fails on his second condition of justification for aims, namely, (*b*).

Finally, grant the skeptic that his aim is nontrivial and reachable. Does he have any argument to show that if the society of scientists were structured along anarchistic lines, then the society would reach the postulated aim? Notice, the argument has to be a strong one. The objectivist view, say, might argue that the society of scientists would have a higher probability, not certainty, of reaching the aim of truth or verisimilitude if the society were structured along the lines of the single best available method. The skeptic finds this claim untenable, for it does not meet his stringent criteria of a justified argument. But the skeptic offers nothing resembling so strong an argument in favor of structuring the society of scientists along his anarchistic lines. Hence, the skeptic's view fails his own third condition, namely, (*c*).[20]

Thus far I have dealt with the skeptic's views on aims; let us turn now to examining his views on theories and methods.

III. Multiply, Multiply, Multiply

Let us return to the convocation and witness possible exchanges of ideas and thoughts that the skeptic would approve of. The following expansion of the tale[21] is meant to illustrate how to proliferate theories, metaphysical systems, and methods.[22]

[20] His failure is compounded. On the skeptic's view, the society of scientists is a vast and varied conglomeration of subgroups, each pursuing its own separate aim. Presumably, the group arrives at its final destination only when each subgroup has arrived at its own ultimate aim. No subgroup, so the skeptic argues, can justify itself; and, therefore, the skeptic's argument in defense of an anarchistic society is to that degree poorer.

[21] See 78–9, this volume.

[22] "Pluralism of theories and metaphysical views is not only important for methodology, it is also an essential part of a humanitarian outlook." Feyerabend, *Against Method*, 52. Or this: "*there can be many different kinds of science . . .* " Thousands of Cuahuila Indians were able to survive in the Southern California desert where today only a few white families exist. Chinese technology flourished on its own steam without the scientific assumptions of the West, and yet was ahead of contemporary Western technology; see Feyerabend, *Against Method* (1988), 3. Or this: "There is not one common sense, there are many (I argued this point with Austin but could not convince him). Nor is there one way of knowing, science; there are many such ways, and before they were ruined by Western civilization they were effective in the sense that they kept people alive and made their existence comprehensible." *Killing Time: The Autobiography of Paul Feyerabend*, 143.

The claim in the one case is, of course, that Cuahuila Indians possessed a different kind of science which enabled them to manage their life (with knowledge of no less than sixty kinds of edible plants) – a science different from the kind of science we

First, the theories. Old, obscure, and radical theories of medicine and anatomy will be revived, such as Avicenna's (ibn Sina's) eleventh-century manuscript *al-Qanun fi al-Tibb* (*The Cannon of Medicine*), Averroes's (ibn-Rushd's) medical encyclopedia *Kuliyat* (composed around 1162), Pietro d'Abano's *Conciliator differentiarum philosophorum et praecipae medicorum* (1300), and not just pathbreakers like Andreas Vesalius's *De humani corporis fabrica* (1543), William Harvey's *De motu cordis* (1628), Georges Cuvier's *Anatomie comparée, Recueil de planches de myologie* (1849–56), and their twentieth-century counterparts. In the field of electricity and magnetism, not only the works of Michael Faraday and James Clerk Maxwell would be studied, but also the theories bound in Franz Aepinus's *Tentamen theoriae electricitatis et magnetismi* (1759), Alessandro Volta's *De vi attractiva* (1769), André-Marie Ampère's *Mémoire sur la théorie mathématique des phénomènes électrodynamique uniquement déduite de l'expérience* (1827), the works of Jean-Antoine Nollet and Giambattista Beccaria, Charles Coulomb and Henry Cavendish. Like those about to be mentioned, these works, would be studied in their entirety (not just those aspects of them that are now regarded as settled science).

In the field of biology and botany the classics will be discussed, of course, such as Carl von Linnaeus's *Systema naturae* (1735) and Charles Darwin's *Origin of Species* (1859), but also René-Antoine Ferchault de Réaumur's six-volume magnum opus *Mémoires pour servir à l'histoire des insects* (1734–42), William Wells's *Essay on Dew* (1818), Patrick Matthew's *On Naval Timber and Arboriculture* (1831), Jacob Agardh's *Theoria systematis plantarum* (1858), Louis Agassiz's four-volume (of the projected ten-volume) work *Contributions to the Natural History of the United States* (1885), as well as the modern synthetic theory of evolution.

In geology, the focus will be on Georgius Agricola's *De re mettalica libri XII* (1558), Johann Lehmann's *Versuch einer Geschichte von Flotz-Geburgen* (1756), and the like, as well as on Charles Lyell's three-volume tract *Principles of Geology* (1830–33) and the twentieth-century researches of geologists such as Tuzo Wilson, F. J. Vine, and D. H. Mathews in the field

possess, which does not enable us to survive there. I leave the reader to distill the claim in the second case, concerning Chinese technology, and then to determine if in either case the claims can shield themselves against scrutiny.

of plate tectonics. In cosmology and astronomy, there will be a similar sort of proliferation: not just Ptolemy and Copernicus, Kepler and Brahe, Caroline Herschel and Hubbel, but also al-Biruni's eleventh-century *Tahdid* and Brahmagupta's *Brahmasphutasiddhanta* (628).

In anthropology, among the theories to be examined will be Anne Robert Jacque Turgot's *Plan for Two Discourses on Universal History* (1750), John Millar's *Observations Concerning the Distinction of Ranks in Society* (1771), J. J. Bachofen's *Mother Right* (1861), as well as Franz Boas's *The Mind of Primitive Man* (1911), Alfred Lewis Kroeber's *Configurations of Culture Growth* (1944), and the works of the French structuralists and cultural materialists.

And what I have sketched for a handful of sciences should also be done for other sciences: cytology, biochemistry, zoogeography, crop science, crystallography, pedology, hydrology, radio physics, geography, oceanography, pathology surgery, population genetics, weed science, marine biology, embryogeny and embryo physiology, Marxian economics, quantum chemistry, forest entomology, silviculture, psychology of learning, physiological psychology, isotope stratigraphy, herpetology, agronomy, sound, ecology, forest entomology, social change, sociology of law, phycology, plant and insect pathology, aquatic plants, galactic astronomy, neuroanatomy, parasitology, toxicology, and so on.

Second, the proliferation of metaphysical systems. Standard fare: Aristotle's *Metaphysics,* Saint Thomas Aquinas's *De Principiis Naturae* (1253), Descartes's *Meditations on First Philosophy* (1641), Baruch Spinoza's *Ethics* (1677), Gottfried Wilhelm Leibniz's *Monadology* (1714), G. W. F. Hegel's *Encyclopedia* (1830), Francis Herbert Bradley's *Appearance and Reality* (1893), Bertrand Russell's *The Philosophy of Logical Atomism* (1918), Ludwig Wittgenstein's *Tractatus Logico-Philosophicus* (1922), Alfred North Whitehead's *Process and Reality* (1929), and Willard van Orman Quine's *Word and Object* (1960).

Not-so-standard fare: Buddhist Mahayana metaphysics, Ali's seventh-century discourses *Nahajul Balagha,* Abu Hamid Muhammad Ghazali's *Tahafut al-Falasifa* (*The Incoherence of Philosophers*), Ulrich of Strasbourg's thirteenth-century fragment *De Summo Bono,* Moses Maimonides's *Dalalat al-Ha'irin* (*Guide of the Perplexed,* 1191), Siger of Brabant's thirteenth-century work *On the Necessity and Contingency of Causes,* Levi Ben Gershon's *Milhamot Adonai* (*Wars of the Lord,* 1560),

Géraud de Cordemoy's seventeenth-century book *Discernment du corps et de l'âme,* Nicolas de Malebranche's *Entretiens sur la métaphysique et sur la religion* (1688), François-Marie Arouet de Voltaire's *Lettres philosophiques* (1734), Schelling's *Ideen zu einer Philosophie der Natur* (1797), Friedrich Daniel Ernst Schleiermacher's *On Religion: Speeches to Its Cultured Despisers* (1799), and Henri Bergson's *Introduction to Metaphysics* (1903).

Third, the proliferation of methods. The convocation opens its doors to methodologists and epistemologists, past and present. The aim is to examine the various theories and metaphysical systems and to develop them in light of vastly different methods. One method may yield advice to develop a theory in one direction; another method may suggest that it be developed in a different direction. If the empirical content of a theory is defined by the set of its potential falsifiers, then each direction in which the theory is explored will increase the empirical content of its rivals, and the exploration of its rivals will in turn increase the empirical content of the theory. Indeed, according to the skeptic, in the absence of rivals, facts cannot be unearthed, and the testability of the theories is reduced. Consequently, if extreme proliferation of theories is the ultimate end, and if the pursuit of a theory is intimately linked to a method, then extreme proliferation of methods is a natural outcome on the skeptic's view.

Another reason for the proliferation of methods is that methods also have to grow and develop, and they can best do that in the same way that theories do. A method improves in the presence of its rivals; methods have to be compared to the history of science, and to each other. In the former case, we examine whether methods have led to successful results; in the latter case, we examine their comparative worth in terms of their ability to solve logical and epistemological problems. But no method, and no theory, should be given up, for we cannot tell how successful it might be in the future. As with theories, so with methods: There should be a constantly increasing pool of mutually incompatible methods offering criteria for demarcating science from pseudo-science, for evaluating competing theories, for selecting theories for practical purposes, and for choosing which theories to pursue for further theoretical research.

Here are some examples. The scientists, as well as the methodologists and epistemologists, in the convocation will read not only standard works in methodology such as John Stuart Mill's *A System of Logic,*

William Whewell's *On the Philosophy of Discovery* and *Philosophy of the Inductive Sciences*, John Herschel's *Preliminary Discourse on the Study of Natural Philosophy*, Charles Sanders Peirce's *Philosophical Writings*, John Dewey's *Experience and Nature*, Carl Hempel's *Scientific Explanation*, Imre Lakatos's *Methodology of Research Programmes*, John Watkins's *Science and Scepticism*, and Hilary Putnam's *Reason, Truth, and History*, but also such books as Sextus Empiricus's *Against the Mathematicians*, Montaigne's *A Defense of Raymond Sebond* (1580), and the works of Parmenides and Archils, student of Anaxagoras and teacher of Socrates.

They will compare the arguments in Karl Popper's *Realism and the Aim of Science* to the antirealist arguments of Luther, Melanchton, and Cardinal Bellarmine, and to the views of the medieval astronomer Bernard of Verdun, who attributed physical reality to eccentrics and epicycles, as well as to the views of Pierre Duhem in *The Aim and Structure of Physical Theory*. They will study Plutarch's *On the Ten Places of Pyrrho* and compare it to the Hippocratic treatise *On Ancient Medicine*, which contained much epistemology. They may even try to form a scientific minisociety, a subgroup, around the methodological works of Pietro d'Abano, Paul of Venice, and Gaetano da Theine. And since these Paduans were deeply imbued in the works of Aristotle, they will read *Posterior Analytics* as well.

Alas, just as the scientists of the past cannot be resurrected, the scientists in the present will not let themselves be associated with all and sundry. Consequently, the skeptic suggests that every thought, idea, and theory – no matter how old and bizarre – be given a chance to be defended. If anyone wishes to defend an anachronistic idea, there should be no reluctance to support the individual and his idea, for "no idea is ever examined in all its ramifications and no view is ever given all the chances it deserves. Theories are abandoned and superseded by more fashionable accounts long before they have had an opportunity to show their virtues."[23] The society of scientists should be brought as close as possible to the society depicted in the convocation, so that it will move swiftly toward producing *The Books of the Sciences*. I have tried to make the tale of the convocation palpable, in order to convey the skeptic's moral, "Multiply, multiply, multiply," so that the reader will not simply dismiss it as a romantic tale but rather conjure

[23] Feyerabend, *Against Method* (1978), 49.

in his imagination what the skeptic's vision might ultimately imply. Perhaps the reader might see therein something noble, something good, something worthy of science, something worthy of man.

IV. Positive Arguments

What are the skeptic's other central arguments? There are positive arguments concerning how the society of scientists should be structured, and there are negative arguments, which criticize alternative methods involving universal rules and standards. Not unmindful of the skeptic's explicit disclaimer that his work is not to be regarded as the mere offering of an alternative set of methodological rules,[24] an overwhelmingly strong case can be made, I think, for the claim that he does have a methodology after all. Indeed, it is from the perspective of his methodology that he criticizes the rival rationalist methods. For this reason, I begin with the skeptic's positive arguments.

The skeptic's method has one supreme rule – which I shall christen *F* – which states, "Anything goes."[25] The other rules of his method are: (*F1*) One ought to entertain a hypothesis that is inconsistent with a generally accepted hypothesis.[26] (*F2*) One ought to develop a hypothesis that is inconsistent with the facts.[27] (*F3*) One ought to accept ad hoc hypotheses.[28] (*F4*) One ought to reject the principle of autonomy.[29] (*F5*) One ought to encourage "massive dogmatism" with respect to a theory.[30] (*F6*) Only strong alternatives should be considered for replacing existing and reasonably successful theories.[31] (*F7*) One ought to proliferate theories.[32] This is surely not an exhaustive list, but it contains some of the more significant rules.

[24] Feyerabend, *Against Method*, 32.
[25] Ibid., 28.
[26] Ibid., 29.
[27] Ibid.
[28] Paul Feyerabend, "Against Method: Outline of an Anarchistic Theory of Knowledge," 38.
[29] Feyerabend, *Against Method*, 38.
[30] Ibid., 298.
[31] Paul Feyerabend, *Realism, Rationalism, and Scientific Method*, 109–10.
[32] "... the principle of proliferation: *Invent, and elaborate theories which are inconsistent with the accepted point of view, even if the latter should happen to be highly confirmed and generally accepted.* Any methodology which adopts the principle will be called a *pluralistic methodology*." Feyerabend, *Realism, Rationalism, and Scientific Method*, 105–6.

I begin by discussing the supreme rule, *F*, namely, "Anything goes." *F* seems to offer much *hope* to a despairing rationalist. After discussing the supreme rule, I shall present objections to it in three parts. In the first part, I shall discuss *F* in relation to itself. In the second part, I shall discuss *F* in relation to the skeptic's methodological rules, (*F1*) through (*F7*). Finally, in the third part, I shall discuss *F* in relation to other methodologies.

To begin with, *F* is not clear. The skeptic expresses very many things with the phrase "Anything goes" – not all of them consistent with one another. In one place he says, "There is only one principle that can be defended under *all* circumstances and in all stages of human development. It is the principle: anything goes."[33] Earlier on, he had said,

> Some of my friends have chided me for elevating a statement such as "anything goes" into a fundamental principle of epistemology. They did not notice that I was joking. Theories of knowledge as I conceive them, *develop*, like anything else. Now there are some people who will accept an epistemology only if it has some stability, or rationality as they are pleased to express themselves. Well, they can have such an epistemology, and "anything goes" will be its only principle.[34]

Then there was the following emendation:

> This is perhaps the place to make some remarks on the meaning of the phrase, 'Anything goes.' Many readers... understand the phrase as recommending chaos and arbitrariness. This is not the case. I do not object to rules, standards, arguments. I only object to rules, standards, arguments *of a certain kind.* I object to rules, standards, arguments which are general, and independent of the situation in which they are applied.[35]

Even this explanation of *F*, or qualification of it, does not damage the rationalist's position – or save that of the skeptic.

In the first place, one may ask whether *F* is "general and independent of the situation in which it is applied"? If the answer is yes, then *F* condemns itself; there is a place after all for normative claims that are general, and the skeptic's system is self-refuting. If the answer is no, then it is not true that anything goes. Thus, *F* either cannot save

33 Feyerabend, *Against Method*, 28.
34 Feyerabend, "Against Method: Outline of an Anarchistic Theory of Knowledge," 105.
35 Paul Feyerabend, "Logic, Literacy, and Professor Gellner," 387. For a defense of Ernest Gellner, see Watkins, "Feyerabend among Popperians 1948–1978," 51–5.

itself or renders harmless the position of the rationalist who proposes general methodological rules.

In the second place, consider the skeptic's own methodological rules, (*F1*)–(*F7*), in light of his *F* principle. Each of these seven rules can be read as a general rule. But given the skeptic's *F*, let us read them as though they were not general rules. What function, or cash value, can these rules then have? The skeptic implores the scientist, before he decides on a course of action, to pay attention to his actual situation, to his history, to his past experiences in dealing with similar problems, his religious and nonreligious beliefs, his goals and aims, and not to consult some general rules.[36] But then the seven rules tend to lose all force. For the skeptic can no longer claim that one ought to encourage hypotheses that are inconsistent with ones that are well established, or to encourage the proliferation of theories, and so on, since in any specific situation one is warranted to do what one pleases. This is a view – as his readers had feared – that would create and encourage arbitrariness, chaos.

In the third place, consider *F* in relation to other methods. Given *F*, a critical rationalist may well take the following stand: "Since anything goes," imagine him saying, "it would seem that the rationalist's method is as much applicable to *any* given situation as the antirationalist's method. For example, the rationalist might advocate abandoning an ad hoc theory, while the skeptic, to the contrary, would advocate pursuing it. Indeed, since the skeptic expressly forbids legislating for the scientist,[37] and concedes that science done in accordance with strict rules is not only *possible* but also to some extent *successful*,[38] the scientist is doubly justified in opting for the rationalist's method. If, on the other hand, some ground is provided for the scientist to prefer the skeptic's seven rules over the rationalist's method, then the principle of 'Anything goes,' *F*, is not operating."

Let me turn from the discussion of *F* to some of the skeptic's more specific methodological rules. In particular, I would like to focus the inquiry on (*F7*), namely, the principle of proliferation. Two things are worthy of notice: First, (*F1*) and (*F7*) are distinct principles. It is

[36] Feyerabend, *Against Method*, 191–6.
[37] Feyerabend, "Logic, Literacy, and Professor Gellner," 384.
[38] Feyerabend, *Against Method*, 22.

not enough that we have two theories that are contradictories, or contraries, of one another; we must have very many theories, each of which is incompatible with the others. This is what would be accomplished in the convocation of the tale. Second, if the principle of autonomy is false – that is, if "there exist facts which cannot be unearthed except with the help of alternatives to the theory to be tested, and which become unavailable as soon as such alternatives are excluded"[39] – then it *follows* that, if we are interested in testing theories, we must adopt the principle of proliferation. This is one instance of how intimately the principles are linked.

Now, the principle of proliferation plays a cardinal role, second only to the supreme principle, in the skeptic's method. Not only is the rationalist's aim of humanism a guiding force in his work, the principle of proliferation is also suggested on the ground that it would both lead to a "higher stage of consciousness"[40] and provide an invaluable and indispensable aid in the growth of objective knowledge.[41] That is, if we want to make progress with respect to objective knowledge, and if we want to discover what is wrong with the theories that we have, the thing to do is to find facts against them. But we cannot find facts against them unless we find and discover alternative incompatible theories. When we find such theories, we shall discover what is wrong with our present theories, and we shall have moved ahead. These insights have now become part of philosophical lore. Yet, within the confines of the skeptic's view, they raise interesting issues and problems.

First, the skeptic wants to entertain every theory, every fable, and every myth. He does not want *anything* discarded. He wants to keep astrology, folk medicine, voodoo, herbal medicine, moxibustion, biblical stories, witchcraft, and, as he puts it, "the ramblings of madmen."[42] Without questioning the value of any of these alternatives, the skeptic's view loses sight of the rationale behind the idea of proliferation.

The rationale of the principle, as well as my criticism of it, will be better understood in light of the situation the skeptic encountered, especially from the rationalist's camp. The skeptic believed that on

39 Ibid., 39.
40 Feyerabend, "Against Method: Outline of an Anarchistic Theory of Knowledge," 30.
41 Feyerabend, *Against Method*, 46.
42 Ibid., 68.

the monolithic methodological view only one theory was allowed to dominate the scientific scene: the one that was most successful. Only *it* was to be expounded in textbooks, taught, studied, financed for further research, and so on – mostly in the name of economy.[43] He argued – rightly, to my mind – that such a monolithic view is injurious to at least one of our aims, namely, the growth of objective knowledge. So the skeptic offered a counter-monolithic view: If one wants to make progress with respect to scientific knowledge, that is, to *replace* one theory with a better theory, *then* one ought to proliferate theories. The society of scientists ought not to be stuck with only one theory; it ought to develop a few more. The idea behind developing incompatible alternative theories is to find out what is wrong with the successful one and to replace it with a stronger alternative.

Now that the skeptic advocates the development of every view, no matter how "trite," it is unclear that the rationale behind the principle of proliferation is the same – if it has any rationale at all. If the monolithic view was economical for the wrong reason, the pluralistic view is utterly uneconomical without reason.[44]

[43] Ibid., 37.

[44] It is just this that is missed in Paul M. Churchland's rather interesting article, "To Transform the Phenomena: Feyerabend, Proliferation, and Recurrent Neural Networks." Taking as an exemplar Feyerabend's discussion of Brownian motion that resulted in conceptual novelty, Churchland defends Feyerabend's thesis of proliferation. His epistemic sympathies, says Churchland, lie more with the liberal Feyerabend than with the more conservative Thomas Kuhn. He offers two arguments: an old neurocomputational argument and a new one. Very briefly, the old argument was intended to show that if humans are multilayered neural networks that learn under the constraints of experience, it matters greatly with what *initial* configuration of synaptic weights a neural network begins. What learning takes place is *path-sensitive,* and some neural networks can get so bogged down in a theory that they cannot displace that theory.

The new neurocomputational argument was designed to show the value of recurrent networks. The recurrent networks show that hidden-layer neurons receive axonal input from two sources rather than one, the sensory input (as in feed-forward networks). The second source is located higher up in the data-processing hierarchy and provides a *bias,* perceptual or interpretational. The result is that a partial, blurry sensory input can be interpreted in light of information supplied by this second source – the phenomena is called *vector completion.* Clearly, the cognitive task performed by a network is a function of the dynamical state of the network, a state that is more than just its physical structure and current sensory input. Given that state, then, what is mere noise to one network is information to another. The "dynamical system of the human brain can occasionally wander into a new region of its vast activational space, a region in which its existing cognitive capacities are put to entirely

Second, proliferation and progress engender problems for the skeptic's view. On the one hand, the skeptic has maintained that he does not know what is meant by progress,[45] and, on the other hand, he claims that his method will lead to success, no matter how we define progress.[46] Lacking knowledge of what is meant by progress does not entitle one to say that one's method, whatever it is, will lead to progress. The skeptic has often argued against the general, all-purpose methodological rules of the rationalist, claiming that they may be

novel uses." Churchland, "To Transform the Phenomena," 156. If this is correct, says Churchland, then "there is a clear and powerful argument for a policy of trying to bring new and unorthodox theories to bear on any domain where we prize increased understanding." Ibid., 157.

There are at least three things worthy of note. First, from Feyerabend's point of view there should *neither* be restrictions on what synaptic weight configurations are to be used, initially or otherwise, *nor* any restriction on what would count as *global* error minimum (in other words, *maximal* performance). Consequently, the difficulties that Churchland mentions for the old model (which led to the new one) are not difficulties at all from a skeptical point of view.

Second, what is missed in Churchland's two arguments is the proliferation of methods. (*a*) Given that methods evaluate theories, and given that methods ought to be proliferated from Feyerabend's point of view, the foregoing point about endless proliferation of theories follows. (*b*) The neurocomputational model would have to show how scientists learn, develop, and sustain methods and methodological values. This is a project that still needs to be accomplished. Churchland might reply that this is already accomplished in talk about synaptic weights. But these seem to be *given* in the model, old or new, and there is no account of why just *these* synaptic weights came to be espoused by the scientists – or, far more interestingly, why they *ought* to be espoused.

Third, Churchland is entirely correct in his assertion that "a single human individual cannot hope to modify his synaptic weight configuration with the ease with which a computer modeler can manipulate an artificial neural network. An adult human's synaptic weights are not open to causal change of the kind required. Hopping briskly around in weight space, thence to explore different learning trajectories, is impossible advice to an individual human. My original advice, therefore, can be realistic only for *groups* of humans, where distinct persons pursue distinct paths." Ibid., 152. One might say that this is a precise statement of the problem of group rationality from the vantage point of a neurocomputationalist. If my arguments in this chapter are right, the neurocomputationalist can ill afford to purchase Feyerabend's principle of proliferation: Every synaptic weight a human has is a restriction. Ergo, a group of scientists must have every available synaptic weight. This is absurd – undoubtedly Churchland would agree. But then to make his neurocomputational model more interesting, for purposes of epistemology, he will first have to solve the normative problem of group rationality, determine which synaptic weights are permissible and why, and then construct a neurocomputational model to reflect this.

45 Feyerabend, "Logic, Literacy, and Professor Gellner," 382.

46 "*And my thesis is that anarchism helps to achieve progress in any one of the senses one cares to choose.*" Feyerabend, *Against Method*, 27.

anti-humanitarian in their effects. He says that, given the opportunity to adopt a theory that is true but unproductive of human well-being and happiness, it is by no means obvious, as the rationalists are only too prone to assume, that we ought to accept that theory. Truth cannot, and perhaps ought not, be the sole criterion here. Let us grant this, and see where the argument leads. Surely the search for truth and the search for human happiness are different aims, and these aims can, and often do, conflict. But if progress is defined in terms of human happiness by one and is defined in terms of an approach toward truth by another, then it is not clear that, no matter how we define progress, the skeptic's method will be successful.[47]

Finally, consider the matter from the point of view of a theory of method, or meta-methodology. A theory of method lays down the conditions under which a method ought to be accepted. There are three questions with which we are presently concerned. First, are there any meta-methodological rules to which the skeptic subscribes in light of which he evaluates other methods? Second, if so, why should a rival methodologist accept the skeptic's meta-methodological suggestions – whatever these are – in favor of his own? Indeed, given the skeptic's anarchistic views, shouldn't the skeptic encourage rival theories of method as well? Third and finally, does the skeptic's method meet the conditions of acceptability of his own meta-methodology?

The skeptic's theory of method – his meta-methodology – consists of at least the following: (*M1*) One ought to prefer a methodology that is humanitarian.[48] (*M2*) One ought not to prefer a methodology that would have blocked the development of science, historically speaking.[49] (*M3*) One ought not to accept a methodology that appears to *guarantee* success and legislates for all places, all times.[50] Each of these rules – and by no means have I exhausted them – is in some ways suspect or held by rationalists themselves. Producing arguments against some of the suspect rules of the skeptic's meta-methodology is enough to show that no sound argument has been given against the rival rationalists' methods.

[47] See also 73–6, this volume.
[48] Feyerabend, "Against Method: Outline of an Anarchistic Theory of Knowledge," 27–36.
[49] Feyerabend, *Against Method*, 171, 183.
[50] Ibid., 20.

Consider (*M1*). First, one may ask, "What is it about a methodology that is humanitarian that makes it preferable to a nonhumanitarian or even an anti-humanitarian one?" Have we found, after all, some common ground between the skeptic and the rationalists? By his anarchistic lights, shouldn't the skeptic encourage nonhumanitarian methods as well? Since happiness and human welfare are so important to the skeptic, why are these produced only by a humanitarian-oriented method? Or does the skeptic find in this aim some unshakable objective argument such that no parallel arguments can be found when discussing scientific theories and metaphysics? Second, one may ask, "If several rival methods are all humanistic, which one(s) shall we prefer?" Is there any ground for believing that the anarchistic view is the only humanistic methodology?[51]

Next, consider (*M2*). The dogma in philosophy of science is that the history of science should be the chief, if not the only, arbitrator among competing methods. The skeptic agrees. He often argues that, if some rationalist's method – say, Popper's – had been accepted by a group of scientists, science as we know it would have come to a halt, or its growth would have been seriously impeded. The skeptic must assume that science has actually progressed if his criticisms against the rationalists are to count (no matter what the rationalists themselves say).[52] For if there is no progress, the rationalist cannot be faulted by citing history. Now, this assumption that the history of science reveals progress involves at least the following ideas: There were certain competing theories at one time, and scientists rightly decided to follow some in favor of others. Had they followed every alternative available, as the skeptic would have had them do, science would not have progressed or would not have progressed at the rate that it did. If so, history refutes the skeptic, if it refutes the methods of the rationalists. (*M2*), the skeptic's meta-methodological rule, refutes (*F*), the skeptic's supreme rule.

There is a further tacit assumption, namely, that if the history of science had been anything other than what it was, there would have

[51] For example, we shall see in the penultimate chapter of this volume that Putnam explicitly disavows Feyerabend-type relativism, yet Putnam's theory of group rationality is not a whit less humanitarian for that.

[52] Paul Feyerabend, "Logic, Literacy, and Professor Gellner," 383.

been no progress. But this is an unrealistic assumption. If rejected theories, such as Aristarchus's heliocentric theory, had been developed, science perhaps would have progressed far more rapidly than it did. Furthermore, we cannot now envisage what new problems of depth and power a rejected theory would have cast ashore, what effect that would have had on the practicing scientists, and what new theories would have arisen as a result. *Realistic alternative and possible histories must be woven if methodologies are to be judged by history.*

Finally, consider (*M3*), which is related to (*M2*). A view that looks to the history of science as an indispensable arbitrator of methodologies I refer to as a *backward-looking view.* Now, there can also be a *forward-looking view.* This view says in effect, "Let us not be overburdened by history. Let us *experiment* with proposed new methods and see how successful they are. Let us be bold in our conjectures about rationality, as we sometimes are with respect to our theories. Even if we fail, we shall have learned a great deal."[53]

There is nothing in the rationalist's methodology that says that his method guarantees success. It would be extremely odd for a rationalist – like Popper – to claim fallibility with respect to scientific knowledge and infallibility with respect to method. What such a rationalist might say is this. As in the case of science, so in philosophy: We begin with problems. It is in light of these problems that we propose our philosophical conjectures; we test for consistency and reasonableness. If our view solves some of our problems and opens up a rich mine of other deep philosophical issues, then there is a lot to be said for our view. If we find that our conjectures about science and rationality are not fruitful, we should abandon them in favor of other, more profitable conjectures. There is no question of certainty, and our methodological rules are not issued with guarantees. If we claimed the latter, we would be saying in effect that we cannot make progress in philosophy. But we do.

Perhaps these counterarguments against the skeptic indicate that the best way of reading him – his words and practice notwithstanding – is to read him *as if* he were not offering any alternative method at all. There is a story about the ancient skeptic Carneades. When in 155 B.C.

[53] The last few paragraphs couch some of the central themes of Sarkar, *A Theory of Method.*

he went to Rome as an ambassador, he delivered a speech in favor of justice. However, the very next day, much to the dismay of his Roman audience, he delivered a second speech, this time against justice.[54] So with our present skeptic: He first argues in favor of a method (positive arguments); then he argues against all methods, his own included (negative arguments). I now turn to these latter arguments.

V. Negative Arguments

These arguments are directed against alternative methods that purport to lay down universal, all-purpose, ahistorical rules: methods that maintain that no matter what the historical stage at which science finds itself, no matter what the particular heritage of theories, methods, and metaphysical doctrines of the scientists, no matter what domain of science the scientists are engaged in, and no matter what their social, political, and economic surroundings, the rules of the method will enable the scientists to make the right choices concerning which theories to accept or reject for purposes of further research and experimentation. In no context of place, time, or domain of science should these rules be disregarded.

Now, here is the skeptic's central negative argument. The rules of the method, says he, "*describe* the situation in which a scientist finds himself. They do not yet advise him how to proceed."[55] That is, a given scientist facing the problem of which theory to pursue is armed by the method only with key notions that as yet tell him nothing about what he should specifically do, unless the method were to specify a *time limit.* The scientist is armed with various notions, such as "empirically promising," "research material," "empirically otiose," and so on. These notions do not help unless accompanied by some time limit.

Let us suppose, says the skeptic, that a theory is research material. But there are well-known examples of theories in the history of science that were once believed to be research material that ended up being empirically otiose. And there are a fair number of cases where a theory that was once regarded as empirically otiose turned around and was

[54] I have poached this story from David Sedley's "The Motivation of Greek Skepticism," 17.

[55] Feyerabend, *Against Method,* 185.

subsequently regarded as research material. Unless one were given some time limit, the scientist would not know how to proceed. Consider sociobiology. Suppose this theory is deemed empirically otiose. Should the scientist not pursue it? No argument is provided to demonstrate why sociobiology might not be a late bloomer, might not become research material in the long run. If it is a late bloomer, then the method that advises scientists to abandon it would be harmful to the growth and development of science.

But, continues the skeptic, time limits cannot be specified. For what rationale could there be for delaying the pursuit of an empirically otiose theory – a year, say, or two – that is not also a rationale for delaying it for three or four years: so goes the slippery slope argument. "Hence, one cannot *rationally* criticize a scientist who sticks to a degenerating programme and there is no *rational* way of showing that his actions are unreasonable."[56] The skeptic's argument is directed, of course, against *any* method that attempts to give universal, all-purpose, ahistorical rules.

Once again, we notice that the skeptic's standards of justification with respect to theories, as before with respect to aims, are extremely stringent. It is no wonder that the methodologist cannot meet them. Essentially, his arguments show the inadequacy of any inductive argument, and these arguments could just as well have come from David Hume or Nelson Goodman. But whereas Hume takes refuge from his skepticism in his notions of habit, custom, and human propensities, and Goodman seeks refuge in entrenched predicates of our language, the skeptic takes in his stride the conclusion that any theory can be pursued with impunity, since no theory can be strongly justified.

Consider his next argument. My method, claims the skeptic, says "*nothing about risks, or the size of risks.*"[57] Were one to speak of risks, one would have to invoke either the cosmological assumption or the sociological assumption, or both. The cosmological assumption states that nature only rarely permits otiose theories to behave like caterpillars. In other words, nature rarely permits an otiose theory to become research material. The sociological assumption states that institutions only rarely permit otiose theories to survive. The institutions are so

[56] Ibid.
[57] Ibid., footnote 12.

structured that editors refuse to publish, foundations refuse to fund, and scientists are outlawed or isolated from the scientific community who pursue otiose theories. The skeptic concludes that while the cosmological assumption needs to be further investigated, "The sociological assumption, on the other hand, is certainly true – which means in a world in which the cosmological assumption is false, we shall forever be prevented from finding the truth."[58]

Let me replace the sociological assumption with what I shall call the group rationality assumption. The group rationality assumption states that the structure of a society of scientists rarely permits empirically otiose theories to survive. If the group rationality assumption is true, then if we are living in a world in which the cosmological assumption is false, we shall forever be prevented from finding the truth. One can infer: If we are living in a world in which the cosmological assumption is false, and we are to find the truth, then a scientific society should have no structure.

The weight of the skeptic's argument lies on the cosmological assumption being false. If the assumption is true, however, then the scientific society ought not to be structured along the lines of the skeptic's view, for then we shall often be pursuing dead ends. What, then, is the truth value of the cosmological assumption? Solving the problem posed by the cosmological assumption is solving the problem of induction, or one version of it. How does one establish the truth of the claim that nature only rarely permits otiose theories to behave like butterflies? Shall we point to several cases in the history of science – Paracelsian medicine, herbal medicine, phrenology, astrology, palmistry, phlogistic chemistry – and conclude that that establishes the claim? But that is simply an inductive argument. For all the past observed cases of otiose theories failing to turn into research material do not establish the claim that *this* theory before us, currently regarded as empirically otiose, will not be a late bloomer and become research material in the future.

Grant that we have failed to establish the truth of the cosmological assumption; we could not solve the problem of induction. Has the skeptic demonstrated that the cosmological assumption is false? How is he to do that? He can point, perhaps, to several cases in the history

[58] Feyerabend, *Against Method*, 186, footnote 12.

of science that were once regarded as empirically otiose, which later bloomed, and conclude that the cosmological assumption is mistaken. But that is still an inductive argument. All the past observed instances that favor the conclusion that the cosmological assumption is false do not establish that the theories now before us, or all future theories, contemporarily regarded as otiose, will turn into research material. It is evident that the skeptic's demands are too stringent; he fails to meet them himself.

VI. The Route to the Goal

Finally, let us recall, duly adapted, the Williams problem:

> What is the ideal social order (Social Utopia) presupposed by the skeptic's theory of group rationality? Is that Social Utopia pliable and morally justified? What is the scientific society, ideally structuring the practice of scientists (Scientific Utopia), presupposed by that same theory of group rationality of the skeptic? Is that Scientific Utopia realistic and how successful would the scientific practice be in that utopia? Can the Social Utopia lie in cohesion with the Scientific Utopia?[59]

How might the skeptic solve the Williams problem? Given the skeptic's view of a rational society (Scientific Utopia), how is that society to be brought about? On the one hand, the skeptic suggests a way, an action, that would produce a rational scientific society without interfering with the rights of individuals. On the other hand, there is a hint of nonaction, a willingness to let history take its own course and hope for the best.

Consider the second suggestion first. The skeptic says, "The attempt to rearrange science or society with some explicit theories of rationality in mind would disturb the delicate balance of thought, emotion, imagination and the historical conditions under which they are applied and would create chaos, not perfection."[60] But surely the skeptic offers a

[59] See Chapter 1, section III, this volume. The skeptic too is concerned with this issue: "How can a society that gives all traditions equal rights be realized? How can science be removed from the dominant position it now has? What methods, what procedures will be effective, where is the theory to guide these procedures, where is the theory that solves the problems which are bound to arise in our new 'Free Society'?" Feyerabend, *Science in a Free Society*, 9. Feyerabend's description of how that Free Society can arise follows.

[60] Ibid., 7.

theory of group rationality, a theory of how to arrange the structure of the scientific society (Scientific Utopia) so that it would attain the skeptic's aim of a rich multiplicity of mutually incompatible alternatives better than any other structure. However, attempting to implant the skeptic's vision of a rational society would disturb the balance of things, a balance the skeptic warns we should leave alone (producing Social Dystopia?). On his view, it is doubtful if there is *any* historical state, condition, or situation that would warrant interference or change (which is not initiated from within, with everyone's acquiescence) without introducing havoc and imperfection.

Like Francis Bacon earlier, the skeptic needs to introduce a distinction between the epistemic states of a scientific society. Some states are epistemically well enough, and they should be left alone; other states call for action, *internal* or *external.* Unless the skeptic introduces even such a broad distinction as epistemically viable versus epistemically nonviable, one simply does not know how his theory of group rationality is to be utilized. Let us suppose he has such a distinction at hand; immediately questions will arise. For example, is it permissible to introduce external changes in some epistemically abysmal, nonviable states of the scientific society? But will not all external changes, for whatever reason, and whatever the historical circumstances, violate some of the rights of individuals the skeptic is so eager to protect? Consequently, are external changes never permissible?[61] On the other hand, how likely is universal acquiescence so that changes may be permitted?[62]

61 "Of course there will be cases when the state rightfully interferes even with the internal business of the traditions it contains (to prevent the spreading of infections diseases, for example) for as with every rule the rules of democratic relativism have exceptions. The point is that in a democracy the nature and the replacement of the exceptions is determined by specially elected groups of citizens and not by experts, and that these groups will choose a democratic relativism as the basis on which the exceptions are imposed." Paul Feyerabend, "Democracy, Elitism and Scientific Method," in *Knowledge, Science, and Relativism*, 223.

Are infectious diseases to be treated according to the Western scientific tradition or, say, Chinese traditional medicine (see Feyerabend, "How To Defend Society Against Science," in *Knowledge, Science, and Relativism*, 186–7)? *Who* – or members of *which* traditions in the state or society – will appoint the specially elected groups of citizens as monitors and implementers of the exceptions? And why would the others be obliged to obey them?

62 If the skeptic is an incipient rights utilitarian (is he?), then he might not need universal acquiescence. In which case, the Utilitarian Problem of Group Rationality formulated earlier and the skeptical solution sketched may be a bit more than a side exploration.

Even granting that possibility, when are internally introduced changes permissible? The skeptic has no answer unless he can draw the lately mentioned broad distinction. The skeptic resists introducing changes – and not only for fear of introducing chaos, but also because any distinction would favor some states of a scientific society over others, favoring some traditions over others. These questions leave the skeptic's view in an unflattering light, unable to solve at least part of the Williams problem, namely, to what extent, if any, can the Social Utopia lie in cohesion with the Scientific Utopia?[63]

Consider now his first suggestion of the way to the goal. The freedom of individuals is paramount, and so, consequently, is a free society.

> *A free society is a society in which all traditions are given equal rights, equal access to education and other positions of power.* If traditions have advantages only from the point of view of other traditions, then choosing one tradition as a basis of a free society is an arbitrary act that can be justified only by resort to power. A free society thus cannot be based on any particular creed; for example, it cannot be based on rationalism or humanitarian considerations. The basic structure of a free society is a *protective structure*, not an ideology, it functions like an iron railing, not like a conviction.[64]

When will such a free society emerge?

> *A free society will not be imposed but will emerge only where people solving particular problems in a spirit of collaboration introduce protective structures of the kind alluded to.*[65]

There are three notions worth focusing on: *protective structure*, *traditions*, and *rationalism*. To begin with the first notion: The basic structure of a free society is a *protective structure*, says the skeptic. Its function is to protect traditions, all of them.[66] Would such a society produce the results the skeptic wants, namely, that all the traditions will flourish unhampered, each in turn producing theories and problems worthy of science? There is no argument here offering any such assurance.

[63] Notice that throughout I have not raised the following part of the Williams problem, "Is the Social Utopia advocated by the skeptic pliable and morally justified?" Strictly speaking, this question belongs in the domain of political philosophy or social justice; I have simply granted to the skeptic that he can answer that question satisfactorily.

[64] Feyerabend, *Science in a Free Society*, 30.

[65] Ibid.

[66] Recall that the structure of a society of scientists is defined in terms of method; hence, the skeptic must be seen as advocating its fundamental principle – "Anything goes" – as defining that structure.

Why assume that the end result will be one favored by the skeptic? For example, what prevents individuals from joining – without force or coercion – a single tradition? Would such a free, single-tradition society be irrational? The skeptic confuses the solution to the problem of social justice (concerned with according rights and duties and the distribution of scarce resources, education, and power) with the solution to the problem of group rationality (concerned with the viability of the scientific enterprise).

What is one to do in a society in which there is an absence of the protective structure? This issue is linked to the second notion, *traditions*.[67] Are externally imposed changes that bring forth the protective structure, by strong or coercive measures, permissible? Let us suppose that they are permissible and that, as a result, a few hitherto neglected traditions are thereby protected. For that to happen, not only must nonhumanitarian traditions be on the rise, no steps must be taken to contain them. But if the society is bound by nonhumanitarian traditions, if only to protect some traditions from others, this will not do, because these nonhumanitarian traditions will surely block or prevent some humanitarian traditions from flowering. Conversely, if some humanitarian traditions are in place, they will try to ensure that no nonhumanitarian tradition takes root. However, the skeptic's view insists that *every* tradition be preserved, given access to power and the means to propagate it through education. Inevitably, it follows that some traditions can survive only by restraining others, or by the demise of others. I conclude: Even when the protective structure is present, the skeptic is mistaken in assuming that every tradition goes.

This is particularly true of how he treats the tradition of *rationalism*, the last of the three notions. According to the skeptic, rationalism is only one tradition among many, one that by historical circumstance and accident has managed to set itself up as the standard by which other traditions should be measured. This would be, at best, question-begging, says he. A society dominated by a single tradition, such as rationalism, is neither a free society nor one that is conducive to the

[67] Elsewhere, Feyerabend has distinguished between what he calls *primary traditions* and *secondary traditions*; see Feyerabend, "Rationalism, Relativism and Scientific Method," in *Knowledge, Science, and Relativism*, especially 203–4. Since this distinction introduces complexities, but does not substantially affect the arguments that follow, it is left unattended.

growth and development of science. Consequently, we need a different scientific society in which a multiplicity of traditions – theoretical and empirical, historical and abstract [68] – would flourish. What is more, "A society that contains many traditions side by side has much better means of judging each single tradition than a monistic society."[69]

The question is, how can that society judge? It is not just difficult but impossible to understand how any particular tradition is to make out a case for itself that is not based on reason. Distinguish reason as a generic notion from reason as a specific notion. One readily grants, for one readily understands, that specific reasons for thinking and doing a thing may vary from culture to culture, from tradition to tradition. But *what* generic notion opposes the notion of reason? Granting that there is such a notion that opposes the notion of reason, it is extremely difficult to understand how, in a scientific society where two or more such notions or traditions prevailed, scientists could decide which was the better tradition without, at the very least, begging the question. Consequently, there is no alternative to the tradition of rationalism, and all alternative traditions are related to the tradition of rationalism as species are related to genus. Only on such a condition – namely, that the ultimate arbitrator is human reason – is comparison of traditions possible. What the skeptic should at most be advocating is the extreme proliferation of such traditions – and no more.

Finally, consider this remark:

> While the skeptic regards every view as equally good, or as equally bad, or desists from making such judgments altogether, the epistemological anarchist has no compunction to defend the most trite, or the most outrageous statement. While the political or religious anarchist wants to remove a certain form of life, the epistemological anarchist may want to defend it, for he has no everlasting loyalty to, and no everlasting aversion against, any institution or any ideology.[70]

But are the skeptic and the epistemological anarchist not the same?

Suppose, first, that they are not the same. Then there must be some conditions under which a certain view is defensible for good reasons.

[68] Feyerabend, *Problems of Empiricism*, 5–8.
[69] Ibid., 31.
[70] Feyerabend, *Against Method,* 189. Note that in this passage Feyerabend's view is designated by *epistemological anarchist,* not by *skeptic.*

Under other conditions, it is not so defensible. Thus, it is not true that every view is equally good or equally bad under every condition. Then the epistemological anarchist – unlike the skeptic – cannot any longer maintain the thesis that anything goes. Perhaps the claim is that a trite or an outrageous idea may be defended for no reason at all, whereas the skeptic may neither accept nor reject it precisely because no good reasons are found on either side. If this is the only difference left between the two views, it is difficult to discern any virtue in the position of the epistemological anarchist as against the position of the skeptic.

Suppose, second, that the skeptic and the epistemological anarchist are the same. What, exactly, are the similarities between them? A political anarchist does not like the capitalist or the socialist structure of his society; he likes a society unshackled by customs, laws, and traditions, one that allows for maximum freedom for the individual, a maximum freedom that cannot be found under more restricted social structures. He prefers society as envisioned by Godwin, Bakunin, or Kropotkin, but not as envisioned by Aristotle, Smith, or Marx. By parallel argument, the epistemological anarchist would prefer a society of scientists unfettered by any all-purpose rules or methods. Neither the single-method view nor the view of multiple methods is his vision of a rational society. Rather, he envisions a society in which scientists have maximum freedom to do, probe, and play with the universe as they please. This is, one would have surely thought, a plausible parallel between the political anarchist and the epistemological anarchist.

Not so. The epistemological anarchist claims that he has no strong attachment, or strong aversion, to any ideology or institution. Presumably, then, he has no commitment to *any* view pertaining to group rationality. He may find himself disposed at one time to accept the classical view; he will favor a society structured along the lines of a single method. And as for that single method, sometimes he may prefer one method, sometimes another. At other times, he may find himself disposed in favor of the view of multiple methods. Sometimes he may favor a society in which anything is permissible; at other times, like the skeptic he distinguishes himself from, he defends no view at all. The view is puzzling. If the society were composed of scientists who were epistemological anarchists, agreement among them as to how to structure their group would be impossible to come by. This because the epistemological anarchist's view encourages neither cohesion nor cooperation, for each scientist in the society might opt for a different aim and, hence, a different solution to the problem of group rationality. Each scientist may defend now this solution, now that. What is more self-refuting than a view that willingly capitulates to the problem of fragmentation?

We end on a theme with which I began this chapter. In the words of Immanuel Kant, in the *Critique of Pure Reason*,

> And thus the skeptic is the taskmaster of the dogmatic sophist for a healthy critique of the understanding and of reason itself. When he has gotten this far he does not have to fear any further challenge, for he then distinguishes his possession from that which lies entirely outside it, to which he makes no claims and about which he cannot become involved in any controversies. Thus the skeptical procedure is not, to be sure, itself **satisfying** for questions of reason, but it is nevertheless **preparatory** for arousing its caution and showing it fundamental means for securing it in its rightful possessions.[71]

Translation, for our purpose: The skeptic is a taskmaster. He constrains the philosopher of science, particularly one offering a theory of group rationality, to develop a sound critique of the concepts and claims the philosopher would use: the usefulness and value of traditional ultimate aims, the restriction on pursuit of theories, and the legitimacy of methods. We have advanced arguments against the skeptic and fear no further challenge; we have learned to distinguish our real possessions from that which lies entirely outside them. His demands are so stringent that no one, including the skeptic, could meet them. Since we do not attempt to do what cannot be done, we cannot become involved in any dispute with respect to it. While, therefore, the skeptical procedure cannot in itself yield any *satisfying* answer to the questions posed by the problem of group rationality, nonetheless it has been *preparatory* to such an answer by pushing hard against the traditional notions of reason and method, and thereby indicating the measures one ought to take, in developing one's theory of group rationality, to secure these legitimate notions or possessions.

It is time to seek alternatives.

[71] Immanuel Kant, *Critique of Pure Reason*, A769/B797.

5

The Subjectivist View I

Quentin Skinner:

If we first ask about the goals Machiavelli urges us to embrace as citizens, these all turn out to be feminine: wealth, liberty, and civic greatness. If we next ask about the type of community we need to sustain in order to realize these goals, we find that this is seen by contrast as masculine: it must be a *vivere politico*, a *vivere civile*, a *vivere libero* – terms usually translated as a 'free state.' If we go on to ask about the qualities needed to uphold such a community, these again prove to be feminine, including *virtu* itself as well as the attribute of prudence, the greatest of political virtues. But if we ask finally about the institutions and arrangements we need to establish in order to promote civic virtues, we find that these – the most basic and shaping elements – are entirely masculine: not merely the term for the general condition of public life (*lo stato*), but also the words used to describe specific institutions (including the Senate and the various magistracies) and the *ordini* or ordinances that bind the whole edifice together.[1]

If we take heed, there is a lot here for us. Our problem of group rationality is modest – too modest, by comparison to the problem Machiavelli was engaged with: to dream up a grand plan for a great society for a conniving prince. Nonetheless, there are obvious parallels. What in a scientific society are the goals scientists should embrace? Say, truth or verisimilitude. What type of scientific community do we need to sustain in order to realize these goals? Say, a free and open society.

[1] Skinner, "Ms. Machiavelli," 30.

What qualities in scientists are needed to uphold such a community? Say, a passion for knowledge. Finally, what is the basic form or structure of the scientific group or society we need to establish in order to promote these qualities? Say, *that* is our problem.

This and the next chapter will reconstruct and evaluate the subjectivist view that is offered as a solution to that problem. The subjectivist view comes in two sharp varieties. Consequently, in this chapter, I focus on the views of Philip Kitcher, and in the next on the views of Thomas Kuhn. Why I lump their views together will become increasingly clear as these two chapters unfold. Kitcher has claimed that a society of Hobbesian scientists (he might just as well have called it a society of Machiavellian scientists), interested in power, glory, and the like, will produce diversity, multiplicity or proliferation of theories, whereas a society of Rousseauean scientists – as I shall call them – interested in the dispassionate pursuit of truth, will not. I shall try to show that Kitcher's subjectivist view does not quite succeed.

To that end, in the first section I address the difficulty of even sketching the problem of group rationality. Then, in section II, I present Kitcher's analysis and arguments that purport to show why the Rousseauean society of scientists will fail to proliferate theories. In Section III, key terms and the Kitcherian division of scientific labor of a society of ruthless egoists are discussed, and a counterexample, among other things, will show how there is *no* proliferation of theories in the society of Hobbesian scientists, at any rate. In Section IV, I study the arguments proposed in evaluating not methods but theories. Finally, in Section V, I present a few cardinal problems for the Kitcherian model of group rationality, topping it with a paradox. The issue surrounding Social and Scientific Utopias, associated with the Williams problem, will not be discussed until the last section of the next chapter, wherein it will be concluded that Kuhn's as well as Kitcher's subjectivist views lead to fairly unhappy conclusions.

I. Individuals, Group, and Goals

Kitcher claims that in a society of scientists, individual scientists may have two epistemic intentions, "personal" and "impersonal." The personal epistemic intention of a scientist, *X*, is defined as follows: "*X* may have the intention that *X* may achieve some epistemic end (to whatever extent is possible)."[2] The epistemic end is, say, truth or verisimilitude. His impersonal epistemic intention is defined as: "*X* may also have the intention that the community to which *X* belongs, the community of past, present, and future scientists, achieve an epistemic end (to whatever extent is possible)."[3] Kitcher then casts the problem of group rationality in the following terms: "[H]ow would scientists rationally decide to coordinate their efforts if their decisions were dominated by their impersonal epistemic intentions?"[4]

Literally understood, there is no conflict between the individual epistemic end and the group end. When the group as a whole has achieved its end, so has the individual a fortiori. The individual scientist wishes to reach truth or verisimilitude; so does the society of scientists to which he belongs. If nothing distinguishes between the two goals, there is no conflict: The individual should do what is in the best interest of the group, and since the group will be able to accumulate more truths than an individual can by himself, the individual's personal epistemic interests are best served by hitching his wagon to the group. Whatever the group would have him do, he should do.

My preliminary efforts notwithstanding, I find the task of precisely determining the problem of group rationality daunting. Minimally, one has to find a common goal for scientists. Let us suppose a scientist is interested in botany; he will also be interested in the neighboring fields, such as taxonomy, genetics, and horticulture. Such a scientist may have no interest in archaeology, quantum electrodynamics, or market fluctuations. His work is devoted to getting to know as much as he can about botany. Ideally, he would want as many resources and as many workers assigned to the fields he is interested in, so that the knowledge in these fields will increase or will increase at a greater speed, enabling him to know more about botany than otherwise. If the society of scientists is structured accordingly, over his lifetime this scientist will come to know more about botany than he would have known if the society of scientists had been structured differently. But the society of scientists is not best served by catering to the interests

[2] Kitcher, "The Cognitive Division of Labor," 9. For further details of Kitcher's views, see his book *The Advancement of Science*, especially Chapter 8.

[3] Ibid.

[4] Ibid.

of a single scientist, or even to a modest subgroup of scientists joined by a common interest in, say, botany. Therefore, either we shall arrive at an impasse, or – a more significant question by far – we must devise a way to show how the interests of the individual scientist and those of the group can have a common factor on which a theory of group rationality can be founded.[5]

To return to the main story, let

S = the society of scientists
i = the individual scientist
R = a set of theories (research programs, paradigms, etc.)
I_i = the evaluation function of i (this function evaluates the epistemic merit of each member in R; each member in S may have a distinct evaluation function)
G_i = the epistemic goal of scientist i
G_{si} = the epistemic goal of the society as postulated by scientist i
G = the epistemic goal of the society as postulated by every scientist.

Each scientist has a certain epistemic goal for himself, his personal goal; and he also has a certain epistemic goal for the group of scientists as a whole, his impersonal goal. For example, he ideally wants maximum truth for himself: his personal goal. On the other hand, he wants maximum truth for the group as a whole: his impersonal goal. This scientist may be willing to sacrifice achieving maximum truth for himself – say, for example, maximum truth in the domain he is interested in – provided that the society of scientists as a whole is better off, but he is unwilling to settle for less in order that another scientist, or subgroup of scientists, is better off while the group as a whole is not. The key notion in his criterion is truth. Suppose, once again, there is another scientist in the group with a different epistemic goal. Say, he ideally wants maximum aesthetic elegance in his own theories: his personal goal. On the other hand, he wants maximum aesthetic elegance in the theories of the group as a whole: his impersonal goal. Like the other scientist, he too is willing to sacrifice achieving maximum

[5] We have, of course, already encountered one version of this important problem as well as the distinction between personal and impersonal interests in discussing the views of Sen; see Chapter 3, especially, 62–3, as well as Chapter 9, section I, 247, and section II, especially 259–60. Since Kitcher does not even reckon the problem, I forgo any further discussion of it.

aesthetic elegance in his own theories provided the society as a whole benefits, but he is unwilling to settle for less aesthetic elegance in his own theories in order that another scientist, or subgroup of scientists, betters itself as a result, while the group as a whole does not. The central notion in his criterion is aesthetic elegance.

The second scientist will not be satisfied with the state of a group of scientists in which there is maximum truth so long as there are theories that are not aesthetically elegant; by contrast, the first scientist will not be satisfied with the state of a group of scientists in which the theories are aesthetically elegant, but there is no reason to believe that they are true or even verisimilar. There is unlikely to be a solution to the problem of group rationality to which every scientist can subscribe, unless *all* scientists in the group can concur on a minimal common epistemic goal. In other words, there must be a $G =$ the epistemic goal of the society as postulated by every scientist. One can almost hear the skeptic of the previous chapter complain, "If my view was deemed objectionable because it led to fragmentation,[6] what prevents fragmentation in this subjectivist view?"

Kitcher has each member in *S* believe, adopt an appropriate "cognitive attitude" toward, the best available theory in *R*. Generalizing on Kitcher's requirement, I shall say that each member in *S* must adopt a cognitive attitude with respect to each member in *R*: in short, *i* must rank order the members of *R*. Kitcher then argues that there can be a discrepancy between individually rational *(IR)* distribution of attitudes and community optimum (*CO*) distribution of attitudes. The *IR* distribution of attitudes among the members of *S* is the one that results when each scientist adopts an appropriate cognitive attitude, in light of his evaluation function, toward each member in *R*. On the other hand, the *CO* distribution, relative to each individual scientist, is that distribution of attitudes among the members of *S* that would maximize the probability of attaining the impersonal epistemic goal of the scientist in question. Kitcher concludes:

> There is a *CO-IR* discrepancy when there is a distribution of attitudes among the members of *S* which, for each *i*, yields a higher probability of attaining *G*, than does the *IR*-distribution.[7]

[6] See Chapter 4, this volume.
[7] Kitcher, "The Cognitive Division of Labor," 10.

First, since Kitcher had said that the problem of group rationality is one in which scientists rationally decide to coordinate their efforts *as if* their decisions were dominated by their impersonal epistemic intentions, the goal whose probability of attainment is increased by the group is not G_i but G_{si}. This is significant, because we have (as we shall see) two versions of *CO-IR* discrepancy coalesced into one.

Second, this is unnecessarily strong. As long as there is a discrepancy between the *IR* and *CO* distributions of attitudes, with respect to even one scientist, that would be enough. For *that* individual scientist, the society of scientists is not rationally organized. One can generalize Kitcher's view: The degree of *CO-IR* discrepancy varies directly with the number of discrepancies. Thus, there is the smallest degree of *CO-IR* discrepancy when the scientific society is not optimally organized according to the goal of only a single individual scientist; at the other end, there is the largest degree of *CO-IR* discrepancy when the scientific society is not organized according to the goal of even a single scientist in the society. Perhaps the society of scientists might aim more modestly at the smallest realistic degree of *CO-IR* discrepancy.

Third, when we seriously take into account the conflict between the epistemic goals of an individual scientist and the epistemic goals of the society of scientists at large, one finds that Kitcher has omitted the possibility of *CO-IR* discrepancy from the point of view of the individual scientist. In short, when the society of scientists is organized so that individual goals of a scientist are met, the goals of the society of scientists are not met, and vice versa. In Kitcherian language, there is, to be sure, a *CO-IR* discrepancy when there is a distribution of attitudes among the members of S that, for each i, yields a higher probability of attaining G_{si} than does the *IR* distribution. But there is also a *CO-IR* discrepancy when there is a distribution of attitudes among the members of S that, for each i, yields a higher probability of attaining G_i than does the *IR* distribution.

Distinguish among various states of the group of scientists:

S_h = the state in which each individual scientist in the group does what is most rational for him (given the circumstances)

S_j = the state in which the group is structured to meet the personal epistemic goals of a single scientist

S_k = the state in which the group is structured to meet the impersonal epistemic goals of a single scientist

S_l = the state in which the group is structured in the best way for the society as a whole.[8]

Quite clearly, S_h is the worst possible state for the individual scientist (although even this might not be true in certain extreme, unrealistic cases); the collective efforts of the group will yield more even for his own benefit, than he could accrue for himself by operating entirely on his own. S_j is, of course, the ideal arrangement, but no scientist would, or should, have even a remote expectation that the society of scientists will be structured wholly to fulfill his particular interests; and S_k, vague though it is, is an ideal too, which may not satisfy the particular goal of an individual scientist but is the best arrangement he might expect. S_l is vaguer yet; but it is no less a crucial state for Kitcher because it is a state in which a resolution among competing impersonal goals of individual scientists in the group will have been reached.

The problem of group rationality from the perspective of Kitcher's subjectivist view, then, can be defined as follows: What arrangements should individual scientists arrive at in order to form a cohesive society (*how* they arrive at such arrangements is of first philosophic importance)[9] so that, for each scientist, S_l is a better state to be in than S_h?[10] Interestingly enough, there must be a sequence of states even between S_j and S_k that a society of scientists could be in. The task would be to discover which state would satisfy the interests of the group as a whole as well as the individual scientists composing the group.

Kitcher points out that his general classification of the problem underscores the importance of the problem of group rationality: The problem arises in the context of differing goals and evaluation functions as well as changes in them over time. He goes on to add that

> The problems are easier to pose and easier to investigate, however, if we suppose uniformity in both respects: that is, that the G_i are all the same and that there is a single ('objective') evaluation function for each scientist.[11]

[8] Once again, the states of a society of scientists need to be distinguished because, clearly, they play an important role in the problem of group rationality. The epistemic states focused on here are different from the ones that have previously concerned us; see, for example, 101–2.

[9] Kuhn's subjectivist view may have something intriguing to say on this; see the next chapter.

[10] Compare our formulation of the problem with this formulation.

[11] Kitcher, "The Cognitive Division of Labor," 10.

In light of the foregoing discussion, one can go a step further and add that if there is no common goal, there can be no possibility of an optimal solution. Note, too, that given Kitcher's assumption that all scientists in the group have the *same* evaluation function – the same method – his view belongs to the set of those methodologies, subjectivist or otherwise, that peddle single methods, and so shares in their defects as well. This assumption of a single method will lead Kitcher's subjectivist view into an intractable problem.

II. Divisions and Discrepancies

How, in a simple concrete case, can *CO-IR* discrepancies occur? Consider, says Kitcher, that the scientific group wants to investigate the structure of a very important molecule. The society of scientists has two empirical methods at its disposal: Method I involves the use of X-ray crystallography, and Method II involves the building of tinker-toy models. Everyone has agreed (presumably, given the evaluation function) that the probability of discovering the structure of the molecule using Method I is far better than that of discovering it by the use of Method II. Since all the members are utterly rational, says Kitcher, they all line up behind Method I.

Let us assume that there are N workers in the society of scientists. Suppose that associated with each method is a probability function $p_i(n)$, representing the probability that Method i will discover the structure of the molecule if n workers are assigned to it. The *CO* distribution is given by assigning n workers to Method I and $N - n$ workers to Method II, thereby maximizing the probability of discovering the structure of the molecule. Assuming, for the sake of simplicity, that the probability of both methods discovering the structure of the molecule is zero, the probability of discovering the structure of the molecule, with the division of scientific labor into two subgroups, is

$$p_1(n) + p_2(N - n)$$

The task of group rationality has been neatly shown to be the problem of how to maximize this function. The solution depends on the form of the function $p(n)$; Kitcher calls these "return functions" because they measure the probability of reaching the goal of the group, the return on its investment of n workers in a given method. One might suppose that the forms of each function for the two methods would be quite different; one method may be more receptive to the efforts of the workers.

Kitcher offers some useful qualitative results of his investigation. The rate of increase is the greatest when n is small; but perhaps this is not realistic. Perhaps the more realistic view, he says, is that the chance of achieving a solution to the problem – discovering the structure of the molecule – increases only slowly at the beginning, since a critical mass of workers may be needed, and this has not yet been reached. Once that mass is reached, then the chance of the discovery will greatly increase, eventually settling down to, or increasing inappreciably above, some number when the workers have saturated the field.

The division of labor will yield good results, depending on the responsiveness of the methods, under the following circumstances. First, when the intrinsic prospects of both methods can be realized with the available work force, it is better to divide the group effort than to assign the group to work on only one method. Second, if the responsiveness of a method is between 1/2 and 1 – so that the intrinsic prospect of one method can be realized, but not that of both – even then, says Kitcher, it pays to divide labor. Third, if Method I is saturated, and Method II would find itself having a critical mass with an additional worker assigned to it, then one should assign the next worker to Method II. Fourth and finally, when the responsiveness of a method has reached 1, it is always better to assign a worker to the method whose intrinsic prospect is higher than that of the other method.

We arrive at the key question: How, then, should an individual scientist behave? To which method will he assign himself? Kitcher offers three views. On the first view, each agent acts simply in accordance with the intrinsic prospects of a method, and not in accordance with what his fellow workers are doing in the field. If we understand individual rationality in this way, we can easily show the discrepancy between *IR* and *CO*. All individuals will assign themselves to Method I, and none will assign themselves to Method II, since they all have the same evaluation function. There will be no proliferation of empirical methods.

On the second view, an individual scientist must assign himself to a method so as to maximize his chance of following a method that would yield the correct answer. This, says Kitcher, can be interpreted in two ways. (*i*) Each scientist makes a decision about which method to join, ignorant of what the other scientists in the group are doing, so that the task is to choose a method whose probability of yielding the right result is as large as possible. (*ii*) Each scientist knows the current division of labor, and must join that subgroup whose probability of yielding the right answer is greater if he joins that subgroup rather than the other subgroup. On either interpretation, Kitcher concludes, a *CO-IR* discrepancy will arise. It is not difficult to see that in the first case, where the individual scientist is making a decision about which subgroup to join without knowing what his fellow workers in the field are doing, a *CO-IR* discrepancy will arise. But how does that conclusion follow in the second case? There is already an existing, stable division of labor (however arrived at); a scientist (who is not part of that original group?) has to make a decision; if he joins, say, the subgroup of Method I, it cannot follow that everyone will, otherwise we cannot possibly have a stable initial division of labor.

Third, we can simply legislate that an individual scientist will act so as to conform to the norms of group rationality. That is,

> an individual rational agent is a person who chooses so as to belong to a community in which the chances of discovering the correct answer are maximized.[12]

This, says Kitcher, is the altruistic ideal of rationality. Kitcher designates it as altruistic, I take it, because the individual scientist would be willing to forgo working in Method I, which he prefers, to work in Method II, which he prefers less, if doing so would increase the probability of group success. To quote Kitcher:

> Altruistically all rational scientists are those who are prepared to pursue theories that they regard as inferior when, by doing so, they will promote achievement of the goals of their own (and their colleagues') impersonal epistemic achievement. Plainly this raises an even more bloodless ideal of scientific rationality than that criticized by historians and sociologists of science.[13]

First, why bloodless? At least in the old-fashioned tradition – to which, Kitcher says, he does not belong – one wants to endow the

[12] Ibid., 14.
[13] Ibid., 9.

rational agent with enough that is relevant from the perspective of epistemology; other factors are kept out as more obfuscating than casting light on the issue.

Second, why that is altruistic, and not quite selfish, is not wholly perspicuous. Kitcher is relying on some unstated assumptions drawn from social theory. I prefer to spend one hour listening to Elgar instead of spending that hour at a community center, but I spend that hour at the community center anyway; what I have lost, the community has gained; I have been altruistic. But the parallel does not obtain here. What the group of scientists gains, I gain; what it loses, I lose. Therefore, it is in *my* interest to join the subgroup that will increase the chance of group success, which would ipso facto be my success, too. Unless Kitcher writes into the goal of a scientist that the latter wants both truth *and* to be its discoverer, this will make Kitcher's point by a trivial method: redefining a term. This is a strategy Kitcher must resist, since elsewhere in his animadversions about ideal rational agents he adverts:

> An obvious move at this point is to modify the requirements of individual epistemic rationality so that the discrepancies vanish by the magic of redefinition: simply declare that an individual rational agent is a person who chooses so as to belong to a community in which the chances of discovering the correct answer are maximized. I suggest that it is a virtue of the analysis I have been presenting that it forces into the open this altruistic ideal of rationality.[14]

Without intending to do so, this mildly caricatures the normative view.[15] A person (say, a Wall Street afficionado) wishes to maximizes his wealth; or a person in a group that wants to maximize its wealth, wishes to act in such a way that the group's wealth is maximized. This states his goal. It is not obvious, however, *how* this individual can act in order to reach, or to come nearer to reaching, that goal. Complex investment portfolios and strategies may have to be designed in order to reach that goal. In a similar fashion, a person may choose to belong to a

[14] Ibid., 14.

[15] It is something like Leibniz's unjust jibe about the method of Descartes. He said Descartes' rules of method were "like the precepts of some chemists; take what you need and do what you should, and you will get what you want." *Philosophischen Schriften von G. W. Leibniz,* IV, 329.

community in which the chance of discovering the truth is maximized, but it does not follow that *how* he is to proceed is thereby obvious. A complex theory of group rationality is called for.

Suppose we devise an invisible hand explanation. A theory of group rationality is offered in which every scientist acts rationally to serve his own ends without intending deliberately and consciously to serve the end of the group as well. But by a mechanism hitherto unexplained, their collective behavior results in the group being designed rationally and leads to promoting its goal as well. This is, of course, the Individual-to-Group Solution, a type of solution we have already covered before.[16] This is minimally what we should strive for, and only if we fail miserably even in this should we subscribe to Kitcher's devastatingly pessimistic view of human reason. For on his account, individual scientists must act irrationally, in the epistemic sense, in order for the group as a whole to be rationally (from an epistemic point of view) structured or designed.

We are now at a point in the discussion where Kitcher concludes that a Rousseauean society of scientists will fail to divide labor as it should be divided.

III. A Society of Ruthless Egoists

Kitcher says:

> By contrast in a neighboring region, the chemical community was composed of ruthless egoists. Each of the members of this community made decisions rationally, in the sense that actions were chosen to maximize the chances of achieving goals, but the goals were *personal rather than epistemic.*[17]

If Kitcher has succeeded in showing this, it must be ranked as an astonishing achievement: for there would then be no reason for connecting a theory of individual rationality to a theory of group rationality – we could dispense with Hilary Putnam's views on individual rationality;[18] or there would be no reason for showing an epistemic connection

[16] See Chapter 1, section II, this volume.
[17] Kitcher, "The Cognitive Division of Labor," 14; my emphasis.
[18] See the penultimate chapter of this volume.

between individuals acting for their own benefit and unwittingly acting for the benefit of the group as a whole.[19]

Kitcher proposes a prize for anyone discovering the coveted solution to the problem, what is the structure of the molecule? He makes a plausible simplifying assumption that if a method succeeds, then each member in that subgroup has an equal chance of winning the prize. Under those assumptions, "How should we expect the Hobbesian community to distribute its efforts?"[20]

Let us suppose that Nostaw is one of the scientists in the field. The current division of labor in the group is $<n, N-n>$. Let us also assume that n is the larger of the two subgroups. Nostaw, curently working in Method I, has to decide whether he should move to Method II. The switch would be good for Nostaw if it would increase the probability of his winning the prize. The probability of his winning the prize is the probability of someone in subgroup II winning the prize divided by the number of people in that subgroup. Thus, at $<n, N-n>$ it is in Nostaw's interest to switch from Method I to Method II, provided

$$\frac{p_2(N-n+1)}{N-n+1} > \frac{p_1(n)}{n}.$$

The next phase of the argument, according to Kitcher, is to discover points of equilibria, points at which no one in the group is better off switching to another subgroup. Here we are offered a sequence of key concepts: The distribution $<n, N-n>$ is *stable downward* if $p_1(n)/n \geq p_2(N-n+1)/(N-n+l)$, and *unstable downward* otherwise. Nostaw should think in terms of moving to subgroup I (or remain there, if he is already there). Likewise, the distribution $<n, N-n>$ *is stable upward* if $p_2(N-n)/(N-n) \geq p_1(n+1)/(n+1)$, and *unstable upward* otherwise. Nostaw should think in terms of moving to subgroup II (or remain there, if he is already there). The distribution $<n, N-n>$ *is bilaterally stable* just in case it is both stable upward and stable downward. If and when the group of scientists reaches a state

[19] Theorists of social justice may take heart: no need to connect the rationality of individual moral agents (by placing them in an Original Position or whatever) to the chievement of a just society. Note: The issues in the field of group rationality and in social justice are both normative.

[20] Kitcher, "The Cognitive Division of Labor," 15.

that is bilaterally stable, *then* we can expect the community to stay there, and it will be optimally distributed.

Kitcher fears, rightly, that stability is one thing, attainability a different story. The trick is to show *how* the group can reach the state of stability, for it is relatively easy to show that once it reaches it, it will stay there. One more term is introduced: "zone of attraction." For any distribution that is bilaterally stable, the zone of attraction is that set of distributions that will collapse to the bilateral state. More accurately, let us say that $<m, N-m>$ collapses up to $<n, N-n>$ just in case $m<n$, and for each x, $m<x<n$, $<x, N-x>$ is unstable upward. Analogously for collapsing downward: $<m, N-m>$ collapses down to $<n, N-n>$ just in case $m>n$, and for each x, $n<x<m$, $<x, N-x>$ is unstable downward. $<n, N-n>$ *is attainable* if its zone of attraction contains all distributions.[21]

Quite significantly, Kitcher concludes:

> The Hobbesian community might work much better than the high-minded spirits of the last section who failed to divide the labor.... The very factors that are frequently thought of as interfering with the rational pursuit of science – the thirst for fame and fortune, for example – might actually play a constructive role in our community epistemic projects, enabling us, as a group, to do far better than we would have done had we behaved like independent epistemically rational individuals. Or, to draw the moral a bit differently, social institutions within science might take advantage of our personal foibles to channel our efforts toward community goals rather than toward the epistemic ends that we might set for ourselves as individuals.[22]

"But is the possibility genuine?"[23] There's the rub.

Kitcher invites us to consider cases. The simplest is one in which the return function is given by

$$p_i(n) = p_i(1 - e^{-kn}),$$

with k large and $p_1 > p_2$. There is a bilateral stable distribution in the neighborhood of $<n^*, N-n^*>$ with $n^* = p_1 N/p_1 + p_2$. The distribution is not only stable, it is attainable. Furthermore, says Kitcher, if p_1 is only slightly larger than p_2, the division of labor is reasonably

21 Ibid., 15–16.
22 Ibid., 16.
23 Ibid.

close to the division demanded by the *CO* distribution, and the probability of the success of the group corresponds to the probability of the success under the *CO* distribution.

Moral: there are conditions under which the Hobbesians do better than their epistemically pure cousins, even conditions under which they come as close as you please to the ideal.[24]

I see *no* argument in support of this last claim; this is not to contest the fact that the distribution would be stable, *if* attained. How does the society of Hobbesian scientists succeed, *if they do*, where the society of Rousseauean scientists fail? In fact, do they succeed? *How*, furthermore, have the social institutions of scientists, taking account of their thirst for fame and fortune – as well as the psychological mechanisms of the scientists, their vices, ranging from greed to being fraudulent, as well as their perseverance, personal investment, personal and national loyalties, and devotion to political causes[25] – managed to finesse that distribution which the more ideal counterparts were not tempted to adopt?

Kitcher himself fears that the group may get stuck at the suboptimal level, particularly at the extreme, $< N - 0 >$. He says, for example, that if we suppose p_2 is too small or N too large, then there may be no benefit to an individual scientist in abandoning Method I. The bad news, Kitcher says, is that when the society of scientists is too big, self-interest leads the group to the same suboptimal state as individual rationality. On the other hand, he asserts, the good news is that the Hobbesian society of scientists not only does better than its "high-minded cousins," it actually achieves a stable optimal division of cognitive labor. Why don't the social institutions of science succeed then? Would the Rousseauean society of scientists do better under those conditions?

Let us grant Kitcher this: These are only extreme, abnormal circumstances in which there is failure to obtain division of labor; hardly surprising, one might say. But does his theory account for normal circumstances? Kitcher offers a way out of the impasse, appealing only to the scientists' selfish interests. Suppose the society of scientists is divided into clans, where each clan has a sufficient number of

24 Ibid.
25 Ibid., 18.

members, say, q. The head of each clan can arbitrarily move people under his jurisdiction any way he wants, so that it would now be possible for q to move about, making it possible to conduct fruitful research on Method II, which was formerly impossible for lack of enough members.

Moral: a certain amount of local aristocracy – lab directors who can control the allegiances of a number of workers – can enable the community to be more flexible than it would be otherwise.[26]

Kitcher has failed to perceive that the objections that defeated the society of Rousseauean scientists also defeat his own Hobbesian society. Or so I shall now argue. Consider a counterexample: There are several such clans, or labs, in the society of scientists; each lab director has a vested interests (the vested interest of his members are plausibly linked to their lab director's). What makes it plausible for one lab director to move to Method II also makes it plausible for *any* other director to move. What Kitcher needs to show, if his Hobbesian community is to succeed in maintaining the *CO* division of labor, is that only one lab director, or a handful of them, would move and that the others could be persuaded, or have reasons of their own, to stay put. Otherwise, the exodus that brought about the collapse of the division of labor, due to a sole reliance on individual rationality, will bring about its collapse here as well. And then the same consideration would apply as before. Nor does this argument question the assumptions Kitcher adverts to, namely, that members are driven by the desire to win a prize and that they have enough information about probabilities, which then guides them in fulfilling their goal. Indeed, if anything, there is a greater propensity on the part of a lab director to jump ship (from Method I) than there would be in the case of less selfish and less egoistic scientists.

Let us be specific. Suppose that there are 500 labs in the country; each lab is working on the problem of discovering the structure of an important molecule. Nostaw is a lab director. Each lab director, *ex hypothesi*, has the same evaluation function. There are two empirical methods that would enable the group to discover the structure

[26] Ibid., 17.

of the molecule, but only one of them (at most) will do so. Assume that the group is stable downward, and would be bilaterally stable at <370,130>. The current distribution stands at <360,140>. The prize is handsome; fame is assured. Nostaw decides to move his lab from Method II to Method I; with luck, so would nine others, making the group bilaterally stable. There are two alternatives, as before. Either each lab director decides to move, or not, without knowing what the others are going to do; or each has information concerning the current distribution of labor in the group and decides whether the addition of his lab will increase their probability of winning the prize and fame.

Here's the difficulty. Not only Nostaw, given his interest in fame and fortune, but *every* lab director has reason to move to Method I. For they all have the same function, the same information, the same probabilities for winning the prize and fame. What prevents any one of them from moving to Method I? Indeed, given that he and his colleagues are Hobbesians, there is even more compelling reason for them to move than there would be for their high-minded cousins. The Hobbesian society of scientists will bring about the collapse of the society to <500, 0> just as readily as would the society of Rousseauean scientists (if, indeed, they would). What this example is meant to show is that while Kitcher's terms are precise, there are no sustaining arguments to show – what he himself suspected, I fear – that a transition to a more desirable state could, let alone would, be effected. For example, consider his notion of "zone of attraction." It is a distribution of attitudes that could collapse into a bilateral state. Grant it, for the sake of argument. However, the real question is: Under what conditions would scientists in the zone of attraction move? Once the conditions are specified, it is not difficult to show that the collapse will occur.

IV. Theory Choice

Let us next consider how Kitcher approaches the issue of theory choice. This issue is thorny, thinks Kitcher, because of two inherent sources of uncertainty: There is a need to take into account the probability that a theory will improve its *apparent* epistemic status, and also the probability that if it does so, it will be closer to the epistemic goal (say, of truth or verisimilitude).[27] Kitcher stipulates, in what follows, that we shall not bother with this source of uncertainty; we shall assume that we are not misled.[28] Let us grant Kitcher the idealization he needs to present a coherent account; our task is to show whether a parallel idealization in the Rousseauean society of scientists will tell a parallel story.

Suppose that we have two rival, incompatible theories, T_1 and T_2. On the basis of the available evidence, T_1 has the probability q_1 of being true, and T_2 has the probability q_2 of being true, and all the scientists in the society recognize this. Suppose further that $q_1 + q_2 = 1$, and that q_1 is approximately equal to q_2.

The community goal is to arrive at the true theory, the universal acceptance of that theory, to eliminate the problems facing the theory, to develop the theoretical and experimental aspects of the theory, and to apply the theory in practical matters. How might the community proceed in trying to achieve this goal? In two ways, according to Kitcher: (*A*) The community may assign all the scientists to one of the two theories, or (*B*) the community may assign n scientists to T_1 and $N - n$ scientists to T_2 (where $0 < n < N$). Kitcher then considers the value, in terms of epistemic utilities, of these two strategies from the vantage point of a later period in the history of science, which he aptly calls "the time of reckoning." These evaluations follow.

Strategy (*A*): If everyone has been following the true theory, then it is the best of all possibilities; the goals have been achieved, or a great many of them have been achieved, and so the society of scientists can be thought to have gained epistemic utility u_1. On the other hand, if everyone has been working on a false theory, then it is the worst of all possibilities; none of the goals have been achieved, and so on, and the society of scientists can be thought to have squandered its effort and gained epistemic utility $-u_1$. Then there is the concept of a "conclusive

[27] The separation between the evaluation or choice of an empirical method, on the one hand, and theory, on the other, appears artificial. For does not a parallel problem arise in the case of the empirical methods? We adopt a certain empirical method – say, Method I – and as we proceed in our application of the method in an attempt to discover the structure of the very important molecule, it yields certain tentative results. Must we not trust that these tentative results give hope that we are on to the road of discovery – in short, that these are not false positives?

[28] Kitcher, "The Cognitive Division of Labor," 19–20.

state." A state is epistemically conclusive if, after each theory has had a fair chance of being examined, developed, and tested, the rivalry between any two (or more) theories is satisfactorily resolved.[29] This will presumably occur when one of the two theories is able to solve its outstanding problems, passing severe testing and the like, while the other has not (and, presumably, is considered unable to do so).

The expected utility of strategy (*A*) is easily computed:

$$q_1 u_1 - q_2 u_1 \ (A)$$

Next, Kitcher calculates the expected utility of strategy (*B*): We choose the distribution of the labor of scientists between the two theories to be $< n, N - n >$; a conclusive state in favor of T_1 comes about just in case T_1 overcomes the anomalies it began with, while T_2 does not overcome its anomalies, despite being given a fair chance to do so. When a conclusive state is thus reached, it is $u_2 (0 < u_2 < u_1)$; the epistemic utility is zero otherwise. The probability that T_1 overcomes its current problems and anomalies is the probability that T_1 is true multiplied by the probability that T_1 responds to the efforts of n scientists. Let $p*_i(n)$ be the probability that a true theory T_i responds to the assignment of n workers by overcoming its problems; the expected utility is

$$q_1 p^*{}_1(n) u_2 + q_2 p^*{}_2(N - n) u_2 \ (B).$$

Kitcher then justly concludes that the division of labor described under strategy (B) is preferable if

$$(B) > (A).$$

In other words,

$$(q_1 p^*{}_1(n) u_2 + q_2 p^*{}_2(N - n) u_2) > (q_1 u_1 - q_2 u_1).$$

It is difficult, if not impossible, to contest this result: Certainly, no one interested in normative methodology would wish to pick a quarrel with it. The trick is to show how individual scientists, allegedly unable

[29] *Fair chance* and *conclusive state* imply a reasonable time limit that rival theories are allowed. But, as the reader might recall, the skeptic doubts that a time limit can be set, and thus he would be skeptical of the possibility of arriving at a conclusive state. I bring up this issue here not because I agree with the skeptic, but to make it clear that the skeptic needs to be answered if Kitcher's subjectivist view is to move forward.

to divide their labor, on the traditional view, can do so by invoking, or appealing to, their nonepistemic interests. And here we run into considerable and crucial difficulties.

V. Problems and a Paradox

Thus, Kitcher:

> Could nonepistemic incentives operate to bring individual scientists close to the *CO*-distribution? Let us suppose that the important motive is each scientist's desire to be singled out by posterity as an early champion of the accepted theory.[30]

If this is all we are required to suppose, the supposition from the earlier arguments fails. It is the supposition, in particular, that the chance of my being the discoverer is equal to the probability of someone in my group discovering something divided by the number of people in my group. To be an early champion of a theory, one that is going to be later accepted, is not the same as *my* making an important discovery or helping to resolve some crucial theoretical problem, or devising a nice experiment to test the theory. *I* might have failed to do any of this, and yet I can be regarded as one of the champions of the theory, provided *my* subgroup as a whole succeeds. This means that the probability of my succeeding, being regarded as an early champion of the theory, is simply equal to the probability of the theory being true. If so, then as an individual scientist I should prefer a theory with the highest probability of success. Once again, even on the Kitcherian model the subgroup collapses to a division of scientific labor that is the same as $< N, 0 >$. Let the scientist be as selfish as he wants, it is not in his interest to give his labor to a theory with a smaller probability.

Let us then suppose that I would regard myself as an early champion of the theory only if *I* made the discovery. Immediately, the question arises, *why* should I, a scientist, *cooperate*?[31] This question is raised

[30] Kitcher, "The Cognitive Division of Labor," 21.

[31] Battles in the late nineteenth century between two vertebrate paleontologists Edward Drinker Cope (1840–1897) and Othniel Charles Marsh (1831–1899) serve to make the point. Marsh tried various subterfuges to prevent Cope from learning about his fossil sites. At least one scientist, Joseph Leidy, unwittingly caught up in this battle, gave up his paleontological pursuits; see E. H. Colbert, *Dinosaurs: An Illustrated History*,

in order to show that the society of scientists, when modeled under strategy (*B*), is remarkably un-Hobbesian. To cooperate is to share with fellow scientists the provisional results, information, theoretical advances, experimental designs or setups, and ideas; to cooperate is to help a fellow scientist when he finds himself in a cul de sac, and so on. The simple answer to the foregoing question is: No, he should not cooperate. He is after all an opportunist, an egoist, a ruthless, selfish Hobbesian. If he is in it for the money, or the prize, or to be regarded as an early champion of the theory that will later be accepted, he reduces *his* chances of being the winner by lending a helping hand.[32] On the Kitcherian model, it is difficult to perceive what *unites* even a subgroup of scientists, let alone the group as a whole. This is once again the problem of fragmentation, but of a different kind. Before this problem resulted from the scientists not sharing a common aim or goal; now it arises because scientists are more focused on *their* getting at the truth rather than in getting at the truth *simplicter* (a goal they do have in common).

25–6. Even so, Colbert says, "The Cope-Marsh feud was a disgraceful affair, yet since each man was trying to outdo his rival it did result in the discoveries of many paleontological treasures, including the skulls and skeletons of dinosaurs." (See also the pungent opening paragraph of S. J. Gould's "Darwin's Untimely Burial.") To be sure. However, I am loath to see Colbert's statement as implying that if one takes away the battles and subterfuges, science will cease to progress. Or that if two enormously gifted paleontologists had *cooperated*, science would not have progressed more rapidly. If Colbert is right, is he right only for the short term; has he failed to take into account the long-term effects produced by a society of scientists that encourages the traits Cope and Marsh exhibited? Or shall we assume that there simply are no costs worth taking into account? Why is this any different from a case in which two warring nations desperately try to prevent each other from gaining access to data, experimental findings, and provisional theories, lest the one beats the other in the arms race?

[32] Charles Darwin and Alfred Russell Wallace would have been rather unfit members of this Hobbesian community so highly prized by Kitcher. In 1870, Darwin wrote to Wallace: "I hope it is a satisfaction to you to reflect – and very few things in my life have been more satisfactory to me – that we have never felt any jealousy toward each other, though in one sense rivals." See Shermer, *In Darwin's Shadow: The Life and Science of Alfred Russel Wallace*, 148. See also Browne, *Charles Darwin: The Power of Place*, 139–40. Or, given powerful friends, which scientist in the Hobbesian community would say as Darwin did about claiming priority over Wallace: "I would far rather burn my whole book, than that he or any other man should think that I have behaved in a paltry spirit"? See Shermer, *In Darwin's Shadow: The Life and Science of Alfred Russel Wallace*, 119.

Let us next suppose that scientists are not entirely ruthless, selfish egoists after all: They *do* cooperate. When someone wins in the subgroup, it is in significant part due to the cooperative venture of the subgroup as a whole; it is as if each member had won. This means that the probability of anyone in the subgroup winning is equal to the probability of the subgroup winning. This is equal to the probability of the theory being true. Then, since $p_1 > p_2$, not even a watered-down Hobbesian – to the extent that he is a Hobbesian at all – has any reason to stay in subgroup II, since it has less probability of winning the prize, or being the theory that will eventually be accepted as the true theory. Therefore, the distribution of scientific labor will be exactly what Kitcher fears will be the distribution on the Rousseauean view, namely, $< N, 0 >$.

Adopt a completely different perspective. Kitcher rightly captures the differences in our two approaches to the theory of group rationality. He remarks that my "focus is on alternatives in methodology rather than in [*sic*] differences in theories, research programmes, or methods."[33] The latter is Kitcher's approach. While he focuses on theories, he shares implicitly the generally shared view (the exception is the skeptic) that a society of scientists should be founded on a single method. A method gives two types of advice: theoretical advice and, more importantly, heuristic advice. Theoretical advice simply tells us which is the best available theory, given the evidence; by contrast, heuristic advice tells us which theory or theories would have the prospect of being a better theory in the future. One can, then, propose a theory of group rationality that enables each scientist to declare the same theory as the best available theory, but enables them to pursue different theories in view of the heuristic advice.

Now, Kitcher grants the distinction, but does not think much of it. In particular:

> This suggested way of avoiding the discrepancy between individual and collective rationality depends on adopting two principles of rationality, one for belief and one for pursuit. The idea that it is rational for a person to believe the better-supported theory seems, however, to be based on supposing that that person's aim is to achieve true beliefs. . . . In that case, however, it appears that the person should also pursue the better-supported theory, since pursuing a

[33] Kitcher, "The Cognitive Division of Labor," 6.

> doctrine that is false is likely to breed more falsehood. Only if we situate the individual in a society of other epistemic agents... does it begin to appear rational for someone to assign herself to the working of ideas that she (and her colleagues) view as epistemically inferior.[34]

This is very hard to fathom. Perhaps Kitcher wants only a single principle of rationality, one for both belief and pursuit. That reading of Kitcher is suspect, since it is not obviously true – indeed, very likely false – that "pursuing a doctrine that is false is likely to breed more falsehood." Furthermore, Kitcher's own Hobbesian scientists avowedly ignore the better-supported theory and pursue the one less supported. One might counter this by saying that that is just the reason why Kitcher invokes Hobbesian scientists: Were they purely interested in epistemic issues, they would not pursue a theory that is less supported than another currently available theory. Grant this for now. It is difficult to understand, then, how a philosopher-monarch (an organizer, a manager) in charge of directing the course of scientific research should adopt a strategy that puts some scientists to work on a theory with a low probability of success. After all, *he* has no stake in the prize. He is a purist: His epistemic goal is the construction of a complete, true story of the world.[35] Since it is the monarch's – or his surrogate's – giving of rewards that distributes scientific labor, is not the monarch drawing on, and relying upon, the distinction between theoretical and heuristic advice? If we do not grant him that, he has no basis for his decisions. But if we grant him that, what prevents us from granting it to the scientists themselves? At the very least, Kitcher denies the distinction at one level, only to introduce it unannounced at another.

Let us weave these arguments together. Consider a simple case of a society of consisting 1,000 scientists. It has five theories, T_1, T_2, T_3, T_4, and T_5, ranked in decreasing order of probability, and its task is to decide, given what it knows about the probabilities of the five theories, how to distribute the scientific labor over these theories so that the future state of the society is better off epistemically under that distribution than it would be under any other distribution.

[34] Ibid., 8.
[35] Ibid.

TABLE 5.1

Theories	T_1		T_2		T_3		T_4		T_5
Probabilities	p_1	>	p_2	>	p_3	>	p_4	>	p_5
Number of workers	500		300		100		75		25

Let us assume that the ideal distribution is as depicted in Table 5.1.

First, it is very important to repeat here Kitcher's operating supposition:

The problems are easier to pose and easier to investigate, however, if we suppose uniformity in both respects: that is, that the G_i are all the same and that there is a single ('objective') evaluation function for each scientist.[36]

It is in light of this assumption that the Rousseauean society of scientists is condemned. On that assumption, all the scientists in the present society, it is said, would work on T_1; to work on any other theory would be irrational. It follows, therefore, that on the Hobbesian model, if we could succeed in getting the scientists to distribute themselves in the optimum way, then 500 scientists would be rational,[37] and the remaining 500, working on the remaining four theories, would be irrational. A *large* core of scientists on the Kitcherian model are epistemically irrational. One should, surely, avoid a model that perforce makes irrationality so endemic in a society that is supposed to be a model of a rational society.

Kitcher argues that, "[o]nly if we situate the individual in a society of other epistemic agents... does it begin to appear *rational* for someone to assign herself to the working out of ideas that she (and her colleagues) view as epistemically inferior."[38] The rationale for this claim is that "since pursuing a doctrine that is likely to be false is likely to breed more falsehood."[39] If this is true – which is unlikely – each Hobbesian individual in the society should reason thus: 'I ought not to

[36] Ibid., 10.

[37] It is uncertain, I think, that we can take even the epistemic rationality of any one of these 500 scientists for granted unless we know the following counterfactual to be false: If no reward had been offered for confirming T_1, then the scientist in question would not have worked on it, and indeed would have left it to work on an inferior theory that *did* hold out a promise of a reward.

[38] Kitcher, "The Cognitive Division of Labor," 8; my emphasis.

[39] Ibid.

pursue an epistemically inferior theory, for I might end up only adding false theories or consequences to the pool of theories. Let someone else do so."

If Kitcher thinks that so long as one is in a group, it is *rational* to pursue an epistemically inferior theory, then he has given us *no* foundation for such a claim to rationality. More significantly: If anyone in a group who works on an epistemically inferior theory can justify his stance as being rational, then there is no need for incentives, economic or otherwise.

Second, the deepest source of Kitcher's problems lies in his failing to consider the proper subject of rationality. As an analogy, consider the following. John Rawls took the subject of social justice to be not the particular actions and judgments of moral agents in a society, but rather the basic institutions of that society; taking an improper subject of social justice would yield skewed results. In a similar fashion, one must ask, what is the proper subject of group rationality? Kitcher takes it to be the particular scientific theories; I take it to be methodology. As a backdrop, and no more, Kitcher's view shares with almost all other theories of group rationality a fundamental feature, namely, that a society of scientists should be structured along the lines of a single best method. Therefore, it is not surprising, given his focus on theories, that any scientist who does not follow the best theory under the method is acting irrationally. And so Kitcher must devise nonepistemic ways to attract scientists to different competing theories; hence the charge of rampant irrationality in his scheme.

By contrast, if we have multiple methods and a distinction between theoretical and heuristic advice, then one is not trapped into having to say what Kitcher says. What is more revealing of its crucial defect, Kitcher's model allows for *no* way in which a society of scientists can improve its knowledge about methods, and not just theories. A view that structures a society of scientists along the lines of multiple methods, for example, can do that, and do so quite naturally.

Third, taking the Adam Smith approach, one might say that if each individual pursued his own economic welfare, the net result in the society would be the economic welfare of all. Even if this is true in the field of economics, the precise parallel in the domain of science would be: If each individual scientist pursued his own epistemic welfare, the net result in a society of scientists would be the epistemic welfare of all.

But, this is not what Kitcher offers, since such a view is too close to the Rousseauean view. Instead, he offers this: If each individual scientist were to pursue his own *economic* interest (or at least half of them, anyway, in our example), we should assume that it would further the *epistemic* interest of the society as a whole. There is no reason to believe that the envisioned consequences would follow naturally. This is even more implausible if, as is more likely, their economic interests and their epistemic interests do not coincide; and if, in fact, their economic interests were to take dominion over their epistemic interests, there would be more reason to anticipate disaster than success.[40]

Fourth, there is a psychological problem. How effectively would a scientist be able to function who, having calculated the probabilities, does not think that a theory is very much worth pursuing from an epistemic point of view, but pursues it anyway on the basis of a desire for fame, glory, and money? And if the psychological problem is serious enough, then one's rational acceptance cannot be so nicely severed from rational pursuit.

Fifth, recall the old argument from Richard Dawkins's *The Selfish Gene* for an "evolutionarily stable strategy."[41] A species of birds is faced with a dangerous tick. Having oneself groomed for a tick enormously increases one's chance of survival. A bird that grooms whoever needs it is a "sucker"; a bird that receives help in grooming but does not

[40] For how the greed of scientists might get the better of them, the most recent and devastating testimony is that of Richard C. Lewontin, "Doubts about the Human Genome Project." Here was a scientific project worth nearly $300 million, funded, if the Alexander Agassiz Professor of Zoology and Professor of Biology at Harvard University was to be believed, on evidence that lay in the region between thin and nonexistent.

Well, this was written over seven years ago. Some scientists, but still not Lewontin, might now argue that the Human Genome Project as we know it is a flourishing project; so scientists are fallible, too. But the case deserves closer study. Was the evidence really poor, and how things turned out sheer dumb luck? Was the claim about the thinness of evidence rushed? How did social factors play a role – especially in producing a mirage of evidence? How can one tell evidence from socially manufactured evidence? See Richard Lewontin, *It Ain't Necessarily So: The Dream of the Human Genome and Other Illusions.*

[41] R. Dawkins, *The Selfish Gene,* 183–6. Kitcher has expressed indebtedness to the discussion of the evolution of sex ratios – in R. A. Fisher's *The Genetical Theory of Natural Selection,* 158–60, and, particularly, in J. Maynard Smith, *Evolution and the Theory of Games* (some of the very authors to whom Dawkins himself has acknowledged indebtedness) – in proposing a theory of group rationality, or how to distribute scientific labor.

in turn help others is a "cheat"; a bird that does not groom anyone who is a cheat is a "grudger." Dawkins argues, conclusively, that faced with this evolutionary situation in which there are only suckers and cheats, suckers would disappear: The strategy of the sucker is evolutionarily unstable. What would be the result in a Kitcherian society of scientists in which half the population is irrational and the other half rational; in which half the population is following epistemic interests and the other half, at least, is following personal interests; in which some cooperate in sharing data, knowledge, and results, and some do not: Which subgroup would have developed a stable evolutionary strategy? Which one would disappear like the sucker? This is, I think, a far more telling way to raise the question I raised earlier: Why should scientists cooperate?[42] If only *some* do, some of the scientists are suckers – we would have reason to fear that their strategies are not evolutionarily stable; if they *all* do – if all the scientists are suckers – then we would have no reason to think that they are Hobbesians after all; if *none* of them do – if they are all cheats – then would a society of such scientists survive?[43]

[42] In the 1989 edition of his book, Dawkins added a fresh chapter entitled "Nice Guys Finish First." This chapter simplifies and brilliantly presents the key results of Robert M. Axelrod's famous book *The Evolution of Cooperation*, and adds its own arguments and findings. The results are extremely pertinent to the issue that concerns us here. Axelrod and Dawkins have found that some plants and insects (fig trees and fig wasps), fish (hermaphrodite fish, the sea bass), mammals (vampires), and men (even in the midst of war) engage in cooperating to their mutual benefit rather than cheating or defecting. It is also interesting to note that those who succeeded were generally nonenvious, charitable, or forgiving and cooperative: a far cry from the members of the Hobbesian group.

Axelrod ran a competition in which he asked people to submit entries for the best way to handle the iterated Prisoner's Dilemma. He divided the submissions broadly into nice (cooperative) and nasty (noncooperative) strategies. In the first round, there were 15 entries; of these, the top 8 were nice strategies; the remaining 7 nasty strategies were far behind. In the second round, there were 63 entries (both rounds included a random strategy invented by Axelrod himself to serve as a baseline). Of the top 15 in the second round, only one was a nasty strategy; of the bottom 15, only one was a nice strategy. The winner in both rounds was Anatol Rapoport. How different would a society of scientists have to be in order for it to be efficient only if it were noncooperative while everywhere in nature it was more successful to cooperate?

[43] It is when struggling with arguments like Kitcher's, as in this paragraph, that I think we should not have leased out so soon Josiah Royce's arguments and vision in *The Philosophy of Loyalty*. In that work, published in 1908, Royce is concerned primarily with idealistic metaphysics – which I eschew – and with moral philosophy. Only a handful of sentences in that book refer in passing to science and societies of scientists. But

Sixth and finally, a paradox. Either the desire for fame and money of a scientist must *always* take second place to his epistemic aims, or it should not. If the first alternative is true, then Kitcher's theory in its fundamental aim is indistinguishable from standard normative theories, and appealing to baser motives is only so much window dressing. For let a scientist's greed or lust for fame be as powerful as you want, *ex hypothesi*, you can offer him no incentive that would make him turn away from his basic epistemic aim. If the second alternative is true, then Kitcher's theory gives *no* reason to think that – barring coincidence – a scientific community that functions in giving epistemic aims second place to other personal aims would actually, but unintentionally, succeed in yielding a more rational society, harvesting scientific theories that have a greater degree of verisimilitude than earlier theories.

Let us compare and conclude. On one reading, the skeptic denies that there is anything to the problem of group rationality; hence no answer need be sought. Another reading of the skeptic has him solve the problem by invoking the cardinal rule, "Anything goes." Kitcher steps away from the skeptic's position: Not only does he take the problem of group rationality to be genuine, he offers a solution that lies at the other end from the skeptic's view. Let there be only one method, says Kitcher – what he calls "a single ('objective') evaluation function." This is surely progress, a step in the right direction. But one that incurs heavy costs along the way.

First, Kitcher's statement of the problem of group rationality, unlike our formulation, leaves virtually no role for method. The structure of a

the arguments Royce uses to admonish moral agents can be drawn upon to illustrate the deep basis on which scientists should conduct themselves in their society. Royce's arguments are intended to be generalized. When loyalties are in conflict, one must act decisively and with fidelity; only through loyalty to a cause that supports loyalty itself, says Royce, can one act from the highest principle while being in a community of loyal wills (see 185 and Chapter 3).

Not so sanguine as to think that Royce has provided us with the ultimate solution – for instance, how does one distinguish one's loyalty to a cause that is worth supporting in a community from a loyalty to a cause that is not? – I nevertheless believe that investigating the problem of group rationality, which seeks a deeper underlying commitment to a cause greater than oneself, and speaking in terms of scientific life plans and purposes would provide a rationale that would undercut the need for providing scientific agents with prizes and enticing gifts to get them to do what is rationally needed. Royce's approach, I venture, should show the non-necessity for a Hobbesian society of scientists (actually, it should show a lot more) and the necessity for a Rousseauean one.

society of scientists is defined, and Kitcher's subjectivist argument proceeds, for all practical purposes, in terms of scientific theories. Such a view does not have enough room to make perspicuous what makes for the growth of reason, namely, improvement in methods. Surely, even Kitcher must applaud if "a single ('objective') evaluation function" is replaced by a more effective evaluative function. How does the replacement occur? By what criteria, if any? What contributes not just to its replacement, but to its greater effectiveness? By what mechanism, if any, do scientists replace one evaluative function by another? Thomas Kuhn, as we shall see, speaks in terms of "negotiation." Would nonepistemic interests be relevant in the selection of an evaluative function as well? If not, why not?

Second, in allowing nonepistemic interests to play an all-powerful role – the ineluctable source of his subjectivist view – Kitcher has simply prevented his theory of group rationality from connecting to theories of individual rationality.

Third, given the vastly different non-epistemic interests that scientists can have, and that can motivate their scientific activities, Kitcher's theory of group rationality succumbs, as did the skeptic's view, to the problem of fragmentation.

Fourth, Kitcher rightly wants a society of scientists designed to proliferate plausible theories, not the extreme version of proliferation given by the skeptic's view. By now, the principle of proliferation is a dogma to which every theory of group rationality must pay tribute. For a group of scientists to pursue a single theory, therefore, is simply unacceptable. (Ask, then: How reasonable is it for a group of scientists to evaluate theories using a single method or evaluation function?) However, Kitcher's model fails to deliver what Kitcher wants: It leads to no proliferation.

Fifth, and finally, the paradox that Kitcher's theory faces is serious: Either nonepistemic interests have no genuine role to play (what Kitcher's theory says notwithstanding), or Kitcher's theory must rely on fantastic coincidences to grant it immunity from failure.

6

The Subjectivist View II

The Structure of Scientific Revolutions, by Thomas Samuel Kuhn (1922–1996), descended upon us like a whirlwind;[1] it utterly transformed pivotal issues in the history and philosophy of science, especially in the field of methodology.[2] Just a list of the terms Kuhn introduced – pre-paradigm, paradigm, normal science and puzzle solving, anomaly, crisis, revolutionary change, incommensurability, textbook-derived tradition – tells by now an old and familiar story. I have little to add to that story – not directly, anyway. This is not because that story led to settled conclusions; hardly any book of any significance does that, let alone Kuhn's book. Rather, what I want to do, in this chapter, is to use some of Kuhn's intriguing concepts and theories in order to fashion out of them – something Kuhn himself never did – a theory of group

[1] Kuhn's book went through three editions. The first edition was published in 1962; the second "enlarged edition" appeared in 1970 and contained an addition of eight lines on p. 84 and and a thirty-seven-page "Postscript-1969"; the last edition, consulted here, was published in 1996 and has an added two-page index.

[2] In his introduction to Kuhn's inaugural Robert and Maurine Rothschild Distinguished Lecture, "The Trouble with the Historical Philosophy of Science," delivered on November 19, 1991, Robert Rothschild wrote in the Foreword: "Kuhn's *The Structure of Scientific Revolutions* (published, as it happens, the same year as Oppenheimer's Whidden Lectures) has stood for three decades as the paradigm of a revolution in the field of the history of science." The praise for its impact on the philosophy of science was scarcely less euphoric. Mary Hesse's essay review began "This is an important book" and ended, "Kuhn has at least outlined a new epistemological paradigm which promises to resolve some of the crises currently troubling empiricist philosophy of science. Its consequences will be far reaching." *ISIS* (June 1963), 286–7.

rationality. My hope is that this will lead to a reexamination of his beguiling texts, for this reason: Kuhn's distinctive subjectivist view of group rationality offers a unique – startling, truth to tell – view of the philosophical landscape.

In section I, I turn my attention to Kuhn's subjectivist view and discuss the role of five key values in scientific decision making and the impact of these values on the group of scientists. In Section II, I delineate how the group makes transitions to various stages that are epistemically interesting and how the group effectively distributes the epistemic risks among its individual members. Section III is reserved for analyzing Kuhn's arguments about history, values, and representative groups. Therein I attempt to show that either Kuhn's view presupposes a powerful, time-independent, normative model, or else Kuhn has not told us – those of us interested in the problem of group rationality, at any rate – enough about why history of science is worth studying.[3] With the foregoing as a backdrop about reasons scientists offer, Section IV devolves upon Kuhn's notion of "negotiation." The fresh, speculative question I raise is: In offering a theory of group rationality, how might the notion of negotiation be conceived and connected to scientists' reasons while staying within the Kuhnian framework? The answer to that question, weaving in as it does the theory of John Rawls, may well be revealing, fruitful even. Finally, in section V, I raise the Williams problem: What *kind* of scientists (Scientific Utopia) and society (Social Utopia) will we end up with, if Kuhn's and Kitcher's views of group rationality are implemented? I conclude that scientists would adopt these utopias only half-heartedly, if at all.

I. Values and Individuals

What are the values that guide all of science? In sum, says Kuhn, there are five: accuracy, consistency, scope, simplicity, and fruitfulness.[4] These are the universals that are present in all the scientific

[3] "Thomas Kuhn's *The Structure of Scientific Revolutions* (1962) forever changed our appreciation of the philosophical importance of the history of science." Michael Friedman, "Remarks on the History of Science and the History of Philosophy," 37.

[4] Kuhn, "Objectivity, Value Judgment, and Theory Choice," in *The Essential Tension*, 321–2. See also, Kuhn, "The Trouble with the Historical Philosophy of Science," especially 13–15.

communities, and hence cannot be used to distinguish one community from another. This, let us say, is Kuhn's thin theory of values; it is succeeded by a far richer theory. Let us call that the thick theory of values.[5] That theory argues that each individual scientist brings to bear his own distinctive tradition and subjective preferences when evaluating theories, and that the uniformity of evaluation promised in older epistemologies is thereby torn down. In its place is put a multiplicity of ways of leaning toward a theory or paradigm. Notice the similarities to and contrasts with Kitcher's version of the subjectivist view: In one respect, both have a single method structuring the society of scientists; in Kitcher, the method is left unspecified; in Kuhn, the method revolves around the aforementioned five values. In another respect, Kuhn deftly introduces a fine variation on that method through the subjective preferences of individual scientists in the group. In Kuhn, these subjective preferences at least mimic being epistemic, I believe; in Kitcher, they are avowedly nonepistemic.

Let us expand this first step in the Kuhnian argument. The traditional theories of epistemology, with their maddening emphasis on objectivity and cavalier ignorance of psychological, social, economic, and political factors in the real lives of individual scientists, would guarantee a uniformity of opinion. For, on these traditional theories, regardless of his background, each scientist will factor into his calculation of the worth of the theory certain objective considerations, such as evidence, test results, observations, experiments, consistency, and deducibility, in tandem with the five basic values. Consequently, each scientist, like an ideal Cartesian rational agent, will see the same simplicity, or the same degree of simplicity, in a theory; he will determine each theory as consistent, through logical analysis; he will measure the same degree of fruitfulness from the actual performance of a theory, and so on. Hence, each scientist will arrive at the same degree of confirmation of a theory, an estimate that is solely a function of the aforementioned facts and values objectively considered. If nothing different is put in, nothing different comes out in the evaluation.

By contrast, the thick theory of values regards the thin theory as simply a common starting point from which different scientists, owing to

[5] In threadbare parallel to John Rawls's distinction between a thin and full theory of good, or Clifford Geertz's distinction between thin and thick descriptions of culture.

their different backgrounds, nationalities, interests, and professional upbringings, will take different roads but – please note – very often arrive at the same destination. Thus, one scientist may heavily emphasize the simplicity of a theory over its consistency, and accept a theory for that reason; while another scientist, coming to the field from a different profession, might regard the theory's simplicity as far less important than its scope. It is the scope of the theory that leads him to accept the theory. They both arrive at the same end point from different starting points; they each accept the same theory but for different reasons.

Next, the second step of the argument: *Why* do scientists place such different emphases on the five values? Kuhn's answer is something of a medley of causes and reasons. One scientist might prefer a theory because that theory exhibits a property he is familiar with in the field from which he has come; another scientist may prefer a theory because of an idiosyncracy of his. There is a host of biographical causes that would explain the differences in decisions among individual scientists, and also, presumably, differences in decisions by a single scientist over an interval of time. For Kuhn, these differences in emphasis are vital, but they cannot be explained in terms of methodology alone. One may speak of nationalities, for instance. French scientists prefer formal models; British scientists prefer mechanical models – so said Pierre Duhem – and so the ultimate explanation rests in social, political, and economic causes (see Fig. 6.1).[6]

"The Dividing Line" in the figure demarcates – to use the standard terms – internal history from external history. "Objective considerations" include the Kuhnian values; John Earman and Wesley Salmon expand that list of objective considerations, while staying strictly within the realm of internal history; arguably, these additional considerations enable us to explain internal history not explainable by the Kuhnian values. Thus far we are still in the realm of normative reasons. On the other hand, "subjective considerations" and "institutional considerations" belong quite properly in the realm that would explain external history; presumably, these considerations transport us out of

[6] *If* one were a vulgar Marxist, seeking reduction of all values to economic values, one might find this to be a neat picture. See Richard W. Miller, *Analyzing Marx: Morality, Power and History*.

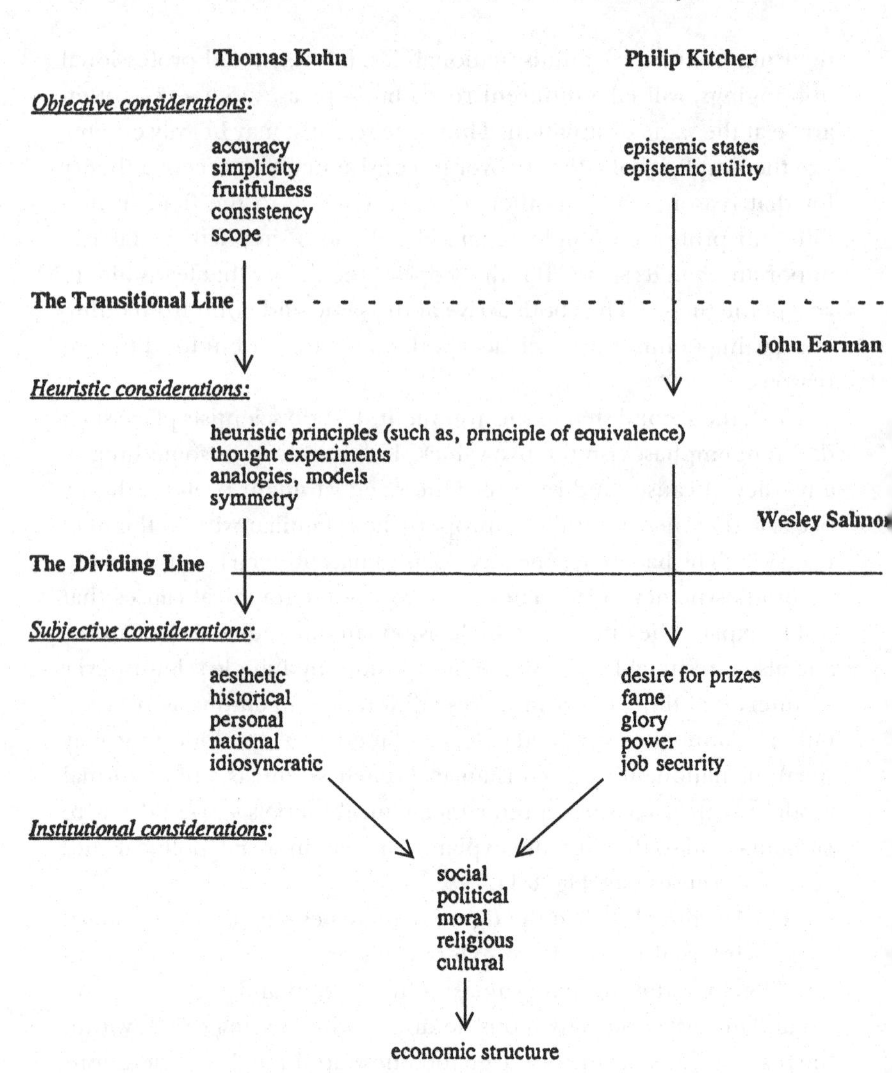

FIGURE 6.1. The origin of theory acceptance.

the normative realm. Kuhn reports, with enormous satisfaction, that Alexandre Koyré – no less – wrote to say that "you have brought the internal and external histories of science, which in the past have been very far apart, together."[7] It follows, from the Kuhnian perspective, that "The Dividing Line" does not exist.

[7] Kuhn, *The Road since Structure*, 286.

Now, here is a puzzle. On this view, no scientist can be regarded as being irrational. For even if he acts on an idiosyncracy, he is rational.

Observation and experience can and must drastically restrict the range of admissible scientific belief, else there would be no science. But they cannot alone determine a particular body of such belief. An apparently *arbitrary element,* compounded of personal and historical accident, is always a formative ingredient of the beliefs espoused by a given scientific community at a given time.[8]

Let us begin with the consequences of the historiographic revolution. One such consequence is that in studying the views of a past scientist, historians have attempted to display the historical integrity of that science in its own time, not in the context of later, modern science. Not Whig history, but history.[9]

They ask, for example, not about the relation of Galileo's views to those of modern science, but rather about the relationship between his views and those of his group, i.e. his teachers, contemporaries and immediate successors in the sciences. Furthermore, they insist upon studying the opinions of that group and other similar ones from the viewpoint – usually very different from that of modern science – that gives those opinions the maximum internal coherence and the closest possible fit to nature.[10]

So far this view is incontrovertible. An objectivist historian of science might applaud that statement and claim that such a view fits in with *his* dogma, too. Where the new historiography collides with the normative view is when it suggests that the historian of science needs to delve deeper in his search for causes and ask what social, political, and economic conditions led Galileo and his teachers, contemporaries, and successors to acquire the beliefs that they did, and that until those conditions are uncovered, the rationality of Galileo's and his contemporaries's beliefs will not have been adequately explained.[11]

[8] Kuhn, *The Structure of Scientific Revolutions,* 4; my emphasis.

[9] For a balanced defense of Whig history, see Howard Margolis, *Paradigms and Barriers,* Chapter 13, "On Whiggishness."

[10] Kuhn, *The Structure of Scientific Revolutions,* 3.

[11] This, of course, hints at the strong program in sociology of knowledge. For Kuhn's relation to those of the strong program, see the rather intriguing remarks of Kuhn in "The Trouble with the Historical Philosophy of Science," especially 8–9. Kuhn there makes every effort to distance himself from the strong program, even calling it "absurd." This is intriguing, I say, because it leads Kuhn, with a hint from Marcello

The normative view suggests that imposed upon the historical model must lie a normative model. No one wishes to maintain that *all* of contemporary science is rational, or that *all* of the beliefs of scientists are rational. Likewise, therefore, one need not assume that all theories of the past were scientific and all decisions and beliefs of past scientists were rational. If so, the historiographic revolution has only made half its journey; the other half will be accomplished when it supplies us with the normative model. This is not to say that all historians of science are interested in the Kuhnian enterprise. An historian may have no interest in pronouncing on the rationality of a scientist, only in explaining *whatever* beliefs a past scientist has held. Once his interests have become less confining, he can safely dispense with the normative model. But surely Kuhn cannot make that escape, or else one would not know how to explain its philosophical relevance. No, Kuhn *is* making claims about how science ought to be done, how scientists ought to make decisions; and, therefore, more than others, Kuhn owes us a normative theory of individual and group rationality. Has he provided us with one?

There is first all "the insufficiency of methodological directives."[12] A scientist embarking upon examining electrical or chemical phenomena, assuming him to be a novice in the field, can take any number of different paths, can draw "any one of a number of incompatible conclusions."[13] Such paths and conclusions are made possible by his prior experience in other fields (such as astronomy), the accidents of his investigations, his own individual make-up, the relevant experiments he first elects to perform, which results strike him hard, and personal and historical accidents and idiosyncracies. Above all, there is an "element of arbitrariness."[14] What is crucial to note, even in this quite incomplete list, is that the objective and subjective considerations are piled together in no particular order. One scientist can reject a theory because it does not have the property he deemed important when he once studied, say, astronomy; another might accept it because

Pera, to alter the very concept of knowledge beyond all recognition. And so I know not how to choose between these two options.

[12] Kuhn, *The Structure of Scientific Revolutions*, 3.

[13] Ibid., 4.

[14] "Yet that element of arbitrariness is present, and it too has an important effect on scientific development." Ibid., 5.

he finds it aesthetically elegant. These are the "essential determinants of scientific development."[15]

> Normal science, for example, often suppresses fundamental novelties because they are necessarily subversive of its basic commitments. Nevertheless, so long as those commitments retain an element of arbitrary, the very nature of normal research ensures that novelty shall not be suppressed for very long.[16]

And that, as we know, will produce the much-desired scientific revolution.

What, then, must restrain the element of arbitrariness? It appears that *nothing* does. Observation and experience appear to be the restraints, but that is only an appearance. Since no paradigm, let alone a theory in the pre-paradigm stage, is without an open problem that is a product of a lack of fit with the data, and furthermore, since observation and experience are inextricably bound to a paradigm, to rules and methods, the restraints seem to be nonoperative; in other words, a scientist can do virtually what he pleases so long as he can hire the element of the arbitrary.

Let me now, briefly, explain Figure 6.1. The arrows exhibit the relationship of dependency. Say, Galileo decides to accept Copernicus's heliocentric theory. The pure historian of science might explain this decision on the basis of objective considerations, such as observations, experiments, the simplicity of the heliocentric model, its scope, its ability to explain novel phenomena, such as the phases of Venus, and so on. That historian might dig up, too, subjective considerations that powerfully influenced Galileo. Such considerations might include ecclesiastical maneuvering, religious sentiments, and so on. At this point, the pure historian of science has no qualms, but the historian of science who has a philosophical point to make may well ask

[15] Ibid., 4.

[16] Ibid., 5. Kuhn's rationale for retaining the element of arbitrariness is: "Nevertheless, so long as those commitments retain an element of the arbitrary, the very nature of normal research ensures that novelty shall not be suppressed for very long." Ibid. But the effect should be quite the opposite, or at least should be slow in coming. If one wants to find a misfit, or a mismatch, between theory and nature, one would be better off with a community that prohibits the element of the arbitrary, which would give rise to different expectations, or similar expectations with different emphases, than a community that permits the element of the arbitrary. In the former case most scientists would expect roughly the same thing, and the likelihood of *all* of them missing the lack of fit or match would be incredibly small.

what bearing, if any, these subjective considerations have on Galileo's rational acceptance of the heliocentric theory. He might ask counterfactual questions like: Had Galileo not been moved by these subjective considerations at all, would he have accepted the theory? Would he then have been justified? Would he have been less justified? What is the epistemic connection, if any, between Galileo's subjective considerations and his objective considerations? The pure historian might, in the meanwhile, already have moved on to a deeper level.[17] Given Galileo's society, what were the institutional features and forces that shaped Galileo's mind, a mind that accepted the heliocentric theory? Were the same historian a Vulgar Marxist, he would link these institutional considerations to the fundamental economic structure and relations, of Galileo's society, attributing the fact that Galileo and his teachers, contemporaries, and immediate successors thought the way they did to their economic interests, motives, and aims as shaped by the society in which they lived.

The normative historian need not regard these explanations as spurious; he claims no interest in them, since such explanations patently do not enable him to distinguish between good science and bad science, between rational beliefs and irrational beliefs. However, once the normative historian of science has a normative model – a model consistent with the historiographic revolution, namely, to seek a theory of individual rationality that will do justice to a scientist's core beliefs and show maximal coherence among them – he might then seek the results of the pure historian of science. The normative historian of science and the pure historian do not have identical interests, but each may learn from the other. The former may learn what were the real historical constraints operating on the individual and the group he wishes to understand; this knowledge can constrain his verdict about the scientist's and the group's rationality. The latter may learn which constraints on his explanations are plausible and which are not.

Interposed between objective considerations and subjective considerations is a row of heuristic considerations. For now, this is intended

[17] For instance, he might investigate the influence that his father's keen rivalry in music theory with Gioseffo Zarlino at the informal Academia Camerata had on Galileo's scientific theorizing, as well as Galileo's invitation to address the Florentine Academy on the location, size, and arrangement of Hell as described in Dante's *Inferno*. See Stillman Drake, *Galileo*, 21 and 23–4, respectively.

as a marker for future serious discussion, but I include the row in the diagram for two reasons. First, it has a direct bearing on Kuhnian exegesis; second, it poses interesting philosophical problems in its own right. Let us begin by taking our cue from John Earman:

> In typical cases the scientific community will possess a vast store of relevant experimental and theoretical information. Using that information to inform the redistribution of probabilities over the competing theories on the occasion of the introduction of the new theory or theories is a process that, in the strict sense of the term is rational: it cannot be accomplished by some neat formal rules or, to use Kuhn's term, by an algorithm. On the other hand, the process is far from being irrational, since it is informed by reasons. But the reasons, as Kuhn has emphasized, come in the form of persuasions rather than proof.[18]

Earman's use of the term "persuasion" is heavily epistemic, as Kuhn's usage assuredly is not, ambiguity and uncertainty surrounding the term notwithstanding. Otherwise, one can fairly ask, "What have nationalities and idiosyncracies and aesthetic quirks to do with epistemology or theory acceptance, let alone that these should be explained in terms of social, political, and economic causes?" One might argue, for instance, that rhetoric plays a large and important role in persuading, where *reasons* have long since failed to persuade. Thus, from an epistemic point of view, Kuhn's stance is far more daunting than Earman's generous reading allows it to be. I concur with Earman, however, that none of this entails that we have a clear notion of what constitutes a heuristic consideration; we do not as yet have even a systematic taxonomy of these considerations.

So, on my view,[19] the transitional line indicates that reasons are still in play, but not yet in the role of proof – whether these are reasons that we employ in attempting to understand a theory, or aiding in the discovery of a theory, or seeking to make the theory plausible (redistributing the priors), or helping to see the interconnections between theories – it is still an epistemic enterprise.[20] By contrast, the

[18] John Earman, "Carnap, Kuhn and the Philosophy of Scientific Methodology," in Horwich (ed.), *World Changes: Thomas Kuhn and The Nature of Science*, 26.

[19] And largely on Earman's view as well, I believe.

[20] Earman's views suggest an interesting investigation that promises a handsome payoff. For example, in making a transition from one theory to another, either ordinary or revolutionary, Earman speaks in terms of logical spaces. He gives an account of how a decision ought to be made in such a transition. He argues that a common framework

dividing line really distinguishes the considerations that have nothing to do with reasons from considerations that do have links with reasons. Wesley Salmon's fear about unadulterated personalism makes a similar point, and his material criteria for assessing the prior probabilities of a hypothesis, involving as it does simplicity, symmetry, and analogy, reflect heuristic considerations.[21]

Kuhn is, I think, ultimately unconcerned about individual rationality, so long as the group as a whole is rational.[22] And this is the third and final phase of the argument. There are, say, *n* scientists in the group; add that there is one very significant theory that ought to attract the attention of these *n* scientists, but *how* or *why* they are attracted to it is of little consequence to the group as a whole, so long as a large number of scientists are drawn toward the theory. In the extreme case, where

can be found in which to compare Newtonian mechanics and Einstein's general theory of relativity, thereby undercutting the need for the incommensurability thesis. Now, part of that account makes use of what I am calling here heuristic considerations. Before a decision could be rendered on empirical grounds, Einstein made, says Earman, masterful use of heuristic devices, such as thought experiments and the principle of equivalence. One might ask, then, "What logical space, if any, that is common to both the theories (Newtonian mechanics and Einstein's general theory of relativity) must be presupposed, in which, say, thought experiments or models must be cast, *before* any decision is made to reassign prior probabilities on grounds of plausibility?" If it is discovered that the logical space can be defined solely in epistemic terms, then the transitional line is well placed; but if other considerations, which Kuhn avers are subjective considerations, are no less significant in nudging the prior probabilities in favor of a new theory, then the dividing line is ill-placed and Kuhn's view of theory acceptance is far more radical than Earman suggests. This is another way of arguing that Earman and Kuhn are nicely divided over what multitude of factors are covered under the rubric "persuasion," even where we have no proof. To cite a purple passage not on Earman's list, consider the paragraph on 152–3 in Kuhn, *The Structure of Scientific Revolutions*; carefully examine the list of what Kuhn says constitutes "reasons" – Kuhn's word – for accepting a new theory and see if they have any discernible epistemic connections. Some of those "reasons" are listed in the foregoing chart. See also Kuhn's *The Essential Tension*, especially 324–5.

21 Wesley C. Salmon, "Rationality and Objectivity in Science: or Tom Kuhn Meets Tom Bayes," in Horwich (ed.), *World Changes: Thomas Kuhn and The Nature of Science*, especially 182–3 and 185–6.

22 Kuhn says, "In the absence of criteria able to dictate the choice of each individual, I argued, we do well to trust the collective judgment of scientists trained in this way. 'What better criterion could there be,' I asked rhetorically, 'than the decision of the scientific group?'." *The Essential Tension*, 321. Directly connected to this claim are an argument and an analogy that come just a few pages later, on 333–4, that have been unjustly neglected. To what extent this makes the two versions of the subjectivist view almost similar in this respect, I leave it to the reader to judge.

all are adhering to the paradigm irrationally, the group as a whole has nevertheless not suffered; in the more usual cases, a few aberrant scientists – who may, for example, continue to sustain an older and now largely discarded tradition, such as "romantic" chemistry[23]– have markedly little effect on the group, so long as most of the members of the group adhere to a single paradigm.

How powerful is Kuhn's ad hoc thick theory of values! First, it can account for the pre-paradigm stage. At this stage, there is no unanimity of opinion because the scientists are making rather different decisions – so different, in fact, that one scientist is led to accept one quasi-paradigm, another to accept a different quasi-paradigm. It might be that one underlines the value of simplicity, which takes him to the first theory; while another scientist underscores the scope of the theory, which leads him to a different theory; a third values accuracy, which leads him to espouse a third quasi-paradigm. Consequently, the thick theory of values can explain why in the pre-paradigm stage theories grow like wild berries.

Second, the thick theory of values can then account for the next stage, the paradigm stage. Notwithstanding their vastly different backgrounds and consequent differences in emphasizing the five key values, the scientists all converge on the same paradigm.[24] One emphasizes simplicity, another scope; yet so successful is the paradigm that it can rush past these different emphases and find something to offer each scientist to entice him to accept the paradigm.[25]

Third, one might think that the thick theory of values would be strained to explain normal science. It would be strained because

[23] "More interesting, however, is the endurance of whole schools in increasing isolation from professional science." Kuhn, *The Structure of Scientific Revolutions*, 19, footnote 11.

[24] Says Kuhn, "What converges as the evidence changes over time need only be the values that individuals compute from their individual algorithms. Conceivably those algorithms themselves also become more alike with time, but the ultimate unanimity of theory choice provides no evidence whatsoever that they do so." *The Essential Tension*, 329.

[25] I share Earman's puzzlement when he says, "But how the community of experts reaches a decision when the individual members differ on the application of shared values is a mystery that to my mind is not adequately resolved by *Structure* or by subsequent writings." Earman, "Carnap, Kuhn and the Philosophy of Scientific Methodology," 20. For Earman's other interesting claims, which are well worth exploring in the context of group rationality, where he argues against a way of forming consensus in science, see 29–31.

different emphases would lead to different evaluations, and different evaluations would lead to different behaviors. Not so; during normal science, a paradigm is not being evaluated or tested. Consequently, these values go into hiding, so to speak; and the differences that they helped to generate in the scientists during non-normal times happily produce no effect, or no pronounced effect, on their work.

Fourth, when normal science is seen to collapse, the rules resurface and scientists once again find themselves in a situation reminiscent of the pre-paradigm stage; there is a multiplicity of theories, or interpretations of the paradigm, and the rules have once again surfaced at just the right time to dominate the scene and make scientists go in different directions. One scientist might regard a problem as a puzzle, another as an anomaly, and a third as a counterinstance. Occasionally it turns out to be a mere puzzle, and the paradigm is once again unquestioned; at other times scientists scatter in all directions.

Fifth, the competition among rules is in full force when the conflict among prospective paradigms, methods, and problems is heightened during the transition period between paradigms. But more of this in the next section.

I conjecture that *if* Kuhn were offering a multiplicity of paradigms as a normal situation – a situation of equilibrium – he would no doubt have invoked his thick theory of values like this:

> When the group of scientists, as a whole, emerges from its pre-paradigm stage at which there are several crude theories into a stage where there are a few good paradigms, it is because not all the paradigms have enough or equal support. Despite their diversities, when comparisons are made of one pre-paradigm to another, and of both to nature, several scientists will find that their respective values do not, and cannot, support the pre-paradigm they are working in. They will therefore leave their pre-paradigm to work under a different theory. Nor will their decisions be unanimous with respect to the theory they ought to be working in: For the rich diversity of their emphases on values will ensure that the group as a whole does not solidify into a single consensual group, but rather scatters into multiple subgroups, subgroups of paradigms.

This sort of group is what Kuhn calls "a heterodox group."[26] Let us note the *sole two* instances where Kuhn claims that there might be a multiplicity of paradigms. First instance: "Each of the schools whose competition characterizes the earlier period is guided by something

[26] Kuhn, *The Structure of Scientific Revolutions*, 164.

much like a paradigm; there are circumstances, though I think them rare, under which two paradigms can co-exist peacefully in the later period."[27] Second instance: "Normally the members of a mature scientific community work from a single paradigm or from a closely related set. In those exceptional cases the groups hold several major paradigms in common."[28]

II. Group Transitions and Risk Distribution

Kuhn asks, "What causes the group to abandon one tradition of normal research in favor of another?"[29] And he answers:

> The scientific enterprise as a whole does from time to time prove useful, open up new territory, display order, and test long accepted belief. Nevertheless, *the individual* engaged on a normal research problem is *almost never doing any one of these things.* Once engaged, his motivation is of a rather different sort. What then challenges him is the conviction that, if only he is skillful enough, he will succeed in solving a puzzle that no one before has solved or solved so well.[30]

The group is emphasized over the individual not only when doing normal science, but also when the transition is made from one paradigm to the next. Kuhn raises the question, "We must therefore ask how conversion is induced and how resisted."[31] Each individual scientist is converted for highly individual reasons, and Kuhn gives a short list of those reasons: the sun worship of Kepler, idiosyncracies of autobiography and personality, nationality and prior reputation of the innovator or his teachers. Not finding these enough, Kuhn concludes:

> Ultimately, therefore, we must learn to ask this question differently. Our concern will not then be with the arguments that in fact convert one or another individual, but rather with the sort of community that always sooner or later re-forms as a single group.[32]

Whatever may be the highly distinctive reasons that led the individuals composing the group to accept (or reject) the paradigm, it is the net results of their decisions on the group, Kuhn thinks, that need to be put in the limelight.

[27] Ibid., ix.
[28] Ibid., 162.
[29] Ibid., 144.
[30] Ibid., 38.
[31] Ibid., 152.
[32] Ibid., 153.

Out of the "Postscript-1969" comes the most significant passage on the problem of group rationality before 1983:

> [I]ndividual variability in the application of shared values may serve functions essential to science. The points at which values must be applied are invariably also those at which risks may be taken. Most anomalies are resolved by normal means; most proposals for new theories do prove to be wrong. If all members of a community responded to each anomaly as a source of crisis or embraced each new theory advanced by a colleague, science would cease. If, on the other hand, no one reacted to anomalies or to brand-new theories in high-risk ways, there would be few or no revolutions. In matters like these the resort to shared values rather than to shared rules governing individual choice may be the community's way of distributing risk and assuring the long-term success of its enterprise.[33]

How, specifically, are the risks distributed? (See Fig. 6.2) In *Phase 1*, there are just a few who have leased their academic lives to a fledgling paradigm, sometimes based on nothing more than aesthetic considerations.

> Though they often attract only a few scientists to a new theory, it is upon those few that its ultimate triumph may depend. If they had not quickly taken it up for highly individual reasons, the new candidate for paradigm might never have been sufficiently developed to attract the allegiance of the scientific community as a whole.[34]

But the few scientists have not accepted the new paradigm because they have found the proof in the pudding. On the contrary,

> The man who embraces a new paradigm at an early stage must often do so in defiance of the evidence provided by problem-solving. He must, that is, have faith that the new paradigm will succeed with the many large problems that confront it, knowing only that the older paradigm has failed with a few. A decision of that kind can only be made on faith.[35]

[33] Ibid., 186. Four years later, Kuhn made a similar claim in *The Essential Tension*, 332. See, too, Kuhn, ten years later, in "Afterwords," 328–9.

Here is a passage, published nine years after the *Structure*, from Bernard Williams: "Moreover, there is the important point for both practical and theoretical enquiries, that each of us is one enquirer among others, and there is a division of epistemic labor, so that what is rational (in this economic or decision-theoretical sense) for *Y* to investigate in detail, it is rational for *Z* to take on *Y*'s say-so." *Descartes: The Project of Pure Inquiry*, 46; see also 47 and 70.

[34] Kuhn, *The Structure of Scientific Revolutions*, 156.

[35] Ibid., 158.

Phase I
The Faith of the Few

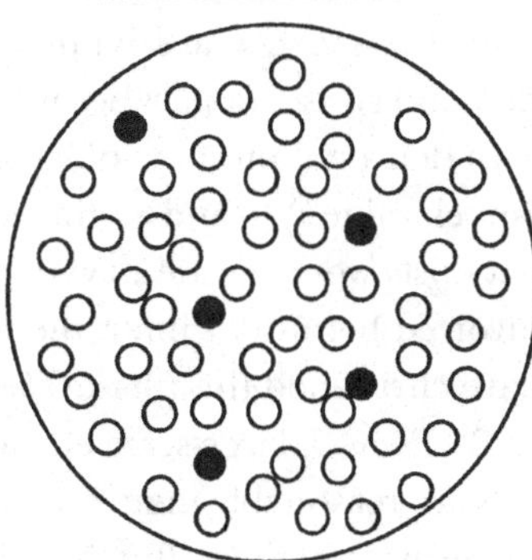

Phase II
The Reasons of the Many

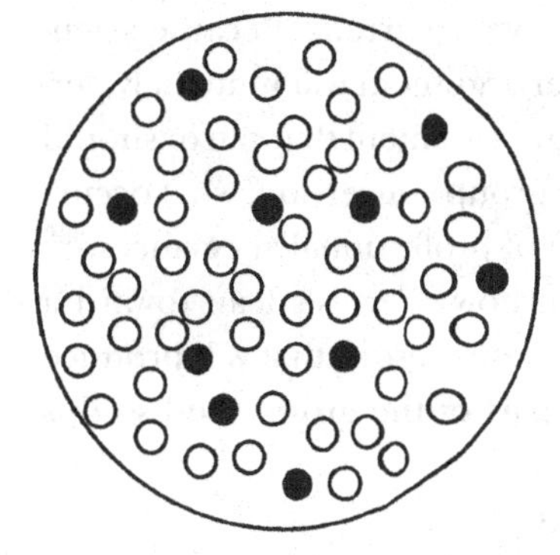

Phase III
The Increasing Shift

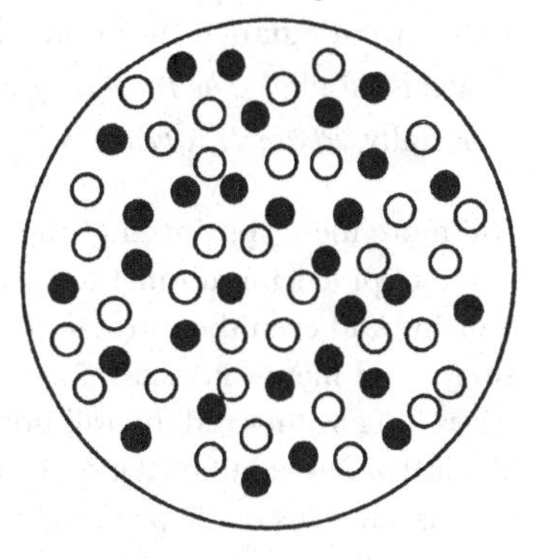

Phase IV
A Few Elderly Holdouts

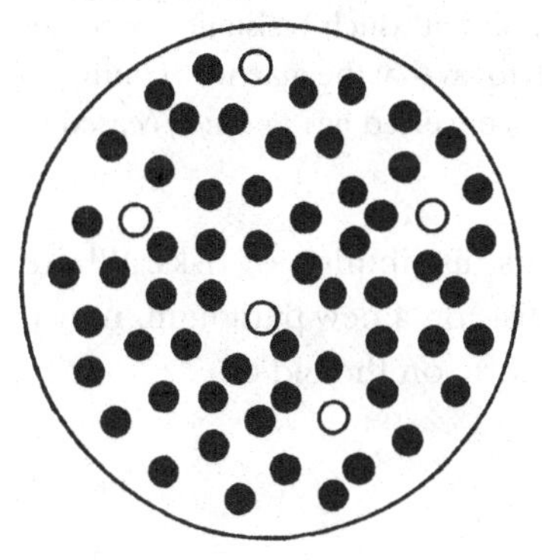

FIGURE 6.2. The group's transitions.

Hence I have titled the first phase *The Faith of the Few.* These few bear the brunt of the risks for the group as a whole. And the risks are large.

In *Phase 2,* several scientists for all kinds of reasons have joined the new enterprise, in part because they see that the new paradigm has at least delivered on some of its promises[36] – there might also have been "shock value"[37] – and so the community is prepared to have as many scientists work on the new paradigm as are proportional to the risks involved because, I infer, the risks now seem slightly smaller. Thus, I have christened this phase *The Reasons of the Many.*

By *Phase 3,* large-scale conversions have occurred. "Because scientists are reasonable men, one or another argument will ultimately persuade many of them. But there is no single argument that can or should persuade them all. Rather than a single group conversion, what occurs is an increasing shift in the distribution of professional allegiances."[38] The thick theory of values will explain how scientists lean toward a theory for distinctive reasons, and it is these reasons that will produce a shift in allegiance in a considerable part of the group. And so this phase is labeled *The Increasing Shift.*

Finally, *Phase 4: A Few Elderly Holdouts.*

> Still more men, convinced of the new view's fruitfulness, will adopt the new mode of practicing normal science, until at last only a few elderly hold-outs remain. And even they, we cannot say, are wrong. Though the historian can always find men – Priestley, for instance – who were unreasonable to resist for as long as they did, he will not find a point at which resistance becomes illogical or unscientific. At most he may wish to say that the man who continues to resist after his whole profession has been converted has *ipso facto* ceased to be a scientist.[39]

Let us say that the group has flipped its distribution of risks: Where before it was willing to risk a few scientists on a new paradigm, now it is willing to risk only as many, but no more, on the old one.

36 Ibid., 154.

37 Ibid., 155.

38 Ibid., 158.

39 Ibid., 159. Two points are worth noting. First, if Priestley is neither illogical nor unscientific, why should he be deemed ipso facto to have ceased to be a scientist? Second, Kuhn failed to see a troubling argument by symmetry. If Lavoisier favored the oxygen hypothesis, when his entire profession had not, should he not then be regarded as having ceased to be scientific, too, the later success of the group notwithstanding?

Now, to talk of risks is to talk of probabilities, but talk of probabilities, in the way required, is prohibited. Hence, risks are impossible to calculate from within the Kuhnian view. Imagine: The group is in the phase where the few are taking the risks. The group would have to calculate how the new theory is likely to evolve in comparison with how the well-installed paradigm will; or the group would have to understand why it is that when a few elderly holdouts maintain their posts, the group does not risk much. Would they not be better employed elsewhere? Or, being old, have they outlived their professional usefulness? But such calculations cannot be made, for they presuppose a common basis for comparing rival paradigms, and there is, on Kuhn's view, no such basis.[40]

Finally, let us even grant that Kuhn can calculate the risks. We now come to the fundamental question: Should the group of scientists distribute the risks in the manner implicitly suggested by Kuhn? That is, should the group of scientists be dominated only by a single paradigm, and should only a few be allowed to take the risks on behalf of the community? Would *it* exhibit the more effective, very specially efficient,

[40] *If* Kuhn is right that there is no pure or neutral observation-language, and paradigms are incommensurable, then I know of no contemporary theory of probability that circumvents Kuhn's cardinal objections and would enable him to calculate the risks for the group. It is no defense of Kuhn to say that he objects in the book only to the old Nagelian version of probability, since Kuhn's arguments are general enough to apply to any version. So Earman's "Carnap, Kuhn, and the Philosophy of Scientific Methodology" and Wesley C. Salmon's "Rationality and Objectivity in Science: or Tom Kuhn Meets Tom Bayes," especially 179–81 and 196–202, concede, in my opinion, far more than they ought. Both these philosophers seek to offer a Bayesian version of Kuhn. Philosophically, this is assuredly a far more interesting task, if a daunting one; I am merely pointing out that in the passages that concern me, neither the arguments of Earman nor those of Salmon should appeal to Kuhn. For this reason: The arguments of the two philosophers are valid only if we deny Kuhn the incommensurability thesis. Earman explicitly and cogently argues against it, and Salmon neglects it for the purposes of his paper. Thus, Earman says, "I deny that there is incommensurability/untranslatability that makes for insuperable difficulties for confirmation or theory choice." "Carnap, Kuhn, and the Philosophy of Scientific Methodology," 17; see also 15 and 18. Salmon says, "I suggested at the outset that an appeal to Bayesian principles could provide some aid in bridging the gap between Hempel's logical-empiricist approach and Kuhn's historical approach. I hope I have offered a convincing case. However that may be, there remain many unresolved issues. For instance, I have not even broached the problem of incommensurability of paradigms or theories." "Rationality and Objectivity in Science: or Tom Kuhn Meets Tom Bayes," 201.

or more rapid rate[41] of movement toward the new paradigm than an alternative proposal? For instance, would the group as a whole be more viable if it permitted a multiplicity of paradigms, at any given time, and assigned to each of them as many individuals as were proportional to the risks involved?

III. History, Values, and Representative Groups

If the social structure is to be invoked to explain why an individual scientist has behaved, or made decisions, in the way in which he has, then it might be enormously interesting to ask what kind of social structure a Kuhnian should, ought to, prefer. Let us narrow the question a bit. One is interested in the question not to determine Kuhnian social and political values, or moral values – in the deeper analysis of the problem, I submit, this would scarcely be without significance – but to determine an answer to this question: *Which social structure, or what kind of society at large* (*Social Utopia*), *would be most beneficial for Kuhnian science* (*Scientific Utopia*)?

Not that Kuhn is unaware of the problem – it is nothing other than the Williams problem, of course – but it manages to earn only a footnote. The passage to which the footnote is attached says:

> If I now assume, in addition, that the group is large enough so that the individual differences distribute on some normal curve, then any argument that justifies the philosopher's choice by rule should be immediately adaptable to my choice by value. A group too small, or a distribution excessively skewed by external historical pressures would, of course, prevent the argument's transfer.

And now the footnote:

> If the group is small, it is more likely that random fluctuations will result in its members' sharing an atypical set of values and therefore making choices different from those that would be made by a larger and more representative group. External environment – intellectual, ideological, or economic – must systematically affect the value system of much larger groups, and the consequences can include difficulties in introducing the scientific enterprises with inimical values or perhaps even the end of that enterprise within societies where it had once flourished. In this area, however, great caution is required. Changes in

[41] Kuhn, *The Structure of Scientific Revolutions*, 164.

the environment where science is practiced can also have fruitful effects on research. Historians often resort, for example, to differences between national environments to explain why particular innovations were initiated and at first disproportionately pursued in particular countries, e.g., Darwinism in Britain, energy conservation in Germany. At present we know substantially nothing about the minimum requisites of the social milieux within which a sciencelike enterprise might flourish.[42]

Combined, these two passages raise a list of nice questions. How large should a group be? What factors should determine its size? How are individual differences in a group to be measured and explained? What is a representative group? What are inimical values, and what (I shall call them by contrast) are catalytic values? Can they be defined independently of the external environment? What is an external environment, and how does it systematically affect the value system? Does a value system systematically affect the external environment? If so, what is the symbiotic relationship between the value system and the external environment? How are changes wrought in the environment to produce, say, inimical values? Would any choice justified by value be immediately adaptable to the philosopher's choice by rule? Is not the philosopher's choice by rule justified even in atypical cases: for instance, when the group is too small or the historical pressures too strong?[43] In what terms is the minimum requisite of the social milieux to be defined?

Consider Graph I in Figure 6.3. The number of individuals in the group, whose differences are represented by the normal distribution curve, is fairly large. The x-axis represents the individual differences; the y-axis represents the number of individuals having the corresponding x number of differences. The bulk of the scientists in the group

[42] Kuhn, "Objectivity, Value Judgment, and Theory Choice," 333, footnote 8.

[43] Perhaps not. Even Rawls had claimed that he expected a certain level of civilization and economic well-being in a society to be in place for "the effective realization of the equal liberties" before his theory of justice could apply to that society. Where those factors were absent, hard choices – contrary to Rawls's theory of justice – would have to be made in order to bring forth a society in which those minimum requirements were satisfied; see Rawls, *A Theory of Justice*, 132. Perhaps, then, Kuhn could argue similarly for a society of scientists. Kuhn would have to tell us what are the minimum requirements (and how that is defined) a society of scientists must satisfy before his subjective theory of group rationality can be applied to that society. Surely this parallel is worth pursuing, but I do not do so here. However, in what follows I do point to some of the problems Kuhn might face.

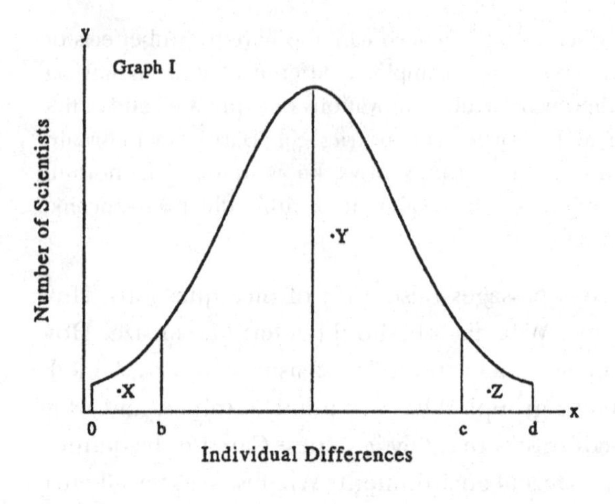

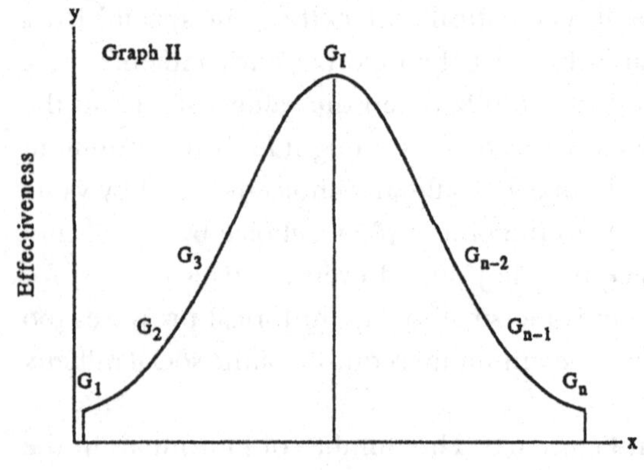

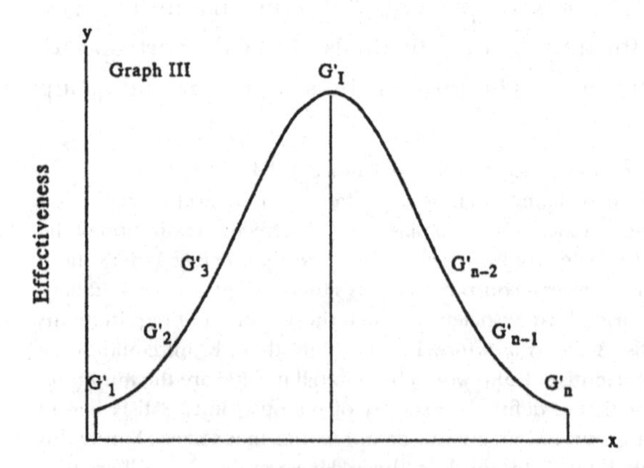

FIGURE 6.3. Number of scientists versus individual differences.

lie between *b* and *c*. Between *a* and *b* lie individuals with extremely small differences, and between *c* and *d* lie individuals with extremely large differences. The number of scientists in these last two cases is quite small; if it were large, the distribution would be skewed. The individual differences represented along the x-axis are differences in how individual scientists tilt on the five values: accuracy, consistency,[44] scope, simplicity, and fruitfulness. As we have seen, one scientist might accentuate simplicity, another scope, a third fruitfulness, and so on. Kuhn suggests the powerful thesis – what is ultimately his central sociological thesis – that these slants, leanings, and accentuations are, in the final analysis, to be explained in terms of the external environment. To quote Kuhn, "External environment – intellectual, ideological, or economic – must systematically affect the value system of much larger groups." This is where Kuhn trespasses the dividing line and traditional philosophers do not. Take away the sociological thesis, and Kuhn's position is not a whit different from the standard view he desires to replace.[45]

Let us, then, suppose that Graph I is about a group of scientists in society S_1 (say, a capitalistic society). In this society, there are scientists *X*, *Y*, and *Z*, among others. *X* belongs to a certain class in S_1's class structure (to be defined in terms of social, political, economic, religious, moral, and intellectual factors), as do those between *a* and *b*. Similarly,

[44] Kuhn regards consistency as utterly crucial for the scientific enterprise and avers that science would alter drastically if scientists did not share this value. "Men did not all paint alike during the periods when representation was a primary value, but the developmental pattern of the plastic arts changed drastically when that value was abandoned. Imagine what would happen in the sciences if consistency ceased to be a primary value." Kuhn, *The Structure of Scientific Revolutions*, 186. Minimally, art would continue to be art even if the value of representation were abandoned; would science even continue to be science if the value of consistency were abandoned? Would it still be science, only strange-looking (as modern art was once thought to be)? Notice, too, that a similar argument is not easily transferred to the other four values.

[45] And yet I would be seriously remiss if I did not report that it was Kuhn himself who claimed to be frequently troubled by the sociological thesis, especially by the labors of those who call themselves Kuhnians in the field of the sociology of science: "I think their viewpoint damagingly mistaken, have been pained to be associated with it, and have for years attributed that association to misunderstanding." "The Trouble with the Historical Philosophy of Science," 3. Consequently, even one who sympathized with Kuhn might well be puzzled and ask, "Just what, in the present context, is the difference?"

Y belongs to a distinct class in S_I's class structure, as do those between *b* and *c*. Likewise for *Z* and those between *c* and *d*. The approximate similarity of (or differences in) their leanings in the judgments they make can then be explained in terms of the classes in the social structure to which they belong. One can construct a similar graph for any other group of scientists in society S_2 (socialistic) or S_3 (Marxist). (Must these class structures be conceived in ideal terms to garner a Social Utopia, or conceived in realistic terms?) Each of these groups of scientists, of course, would constitute a representative group. Each such group would reflect in its individual differences the class structure of the society in which its scientists exist. This, of course, does not begin to tell us which of the three groups is more effective, unless we have an independent – that is, independent of the external environment or class structure – criterion to serve as a measure of effectiveness.

It is true, let us say with Kuhn, that at present we know substantially nothing about the minimum requisites of the social milieux within which a sciencelike enterprise might flourish. But suppose we did. Then we could offer the following, along Kuhnian lines (see Graph II). Consider various groups of scientists G_1, G_2, G_3, ... G_i, ... G_n. Each is large enough and constitutes a representative group. These groups can be ordinally arranged to constitute a normal curve, with G_I, the ideal group, at the peak, and all other groups removed from it, their distance being proportional to their effectiveness relative to G_I, and their effectiveness, in turn, a function of the distance of their social milieux from that of the social milieux of G_I. (Again, must the social milieux be conceived in ideal or realistic terms?) How the effectiveness of a group is to be measured is taken for granted, not explained. We assume that the various groups of scientists are handicapped to various degrees by the inimical values that are held by the scientists in those groups; these inimical values are to be explained in terms of the external environment. What is also left unexplained is the external environment that produces the minimal requirement of values necessary for science to flourish. G_I is the only one to satisfy the minimum requirements of the social milieux within which a sciencelike enterprise might flourish. (Must not there be a criterion of what constitutes science that is independent of a social milieux?) Thus, G_I and G_n would be furthest removed from G_I and would be least effective, and G_{I-1} and

G_{I+1} but a shade away from being a society in which science might flourish.

Next, assume that various groups of scientists G'_1, G'_2, $G'_3, \ldots G'_i, \ldots G'_n$ can be ordinally arranged to constitute a normal curve with G'_1 at the peak and all other groups removed from it, their distance being proportional to their effectiveness relative to G'_I (see Graph III). G'_I, the ideal group in this class, is the group that does best among the groups that *do* satisfy the minimum requirements of the social milieux within which a sciencelike enterprise might flourish. Thus, G'_I and G'_n would be furthest removed from G'_I and would be least effective, and G'_{I-1} and G'_{I+1} but a shade away from being a society in which science would be at its best.

First, if Kuhn's argument is to hold up, even in principle, he must give us an account of the nature of science – shall we say a normative account? – *before* we can judge whether a particular social milieux is going to harm science and its research or help it. But if this cannot be done without invoking the notion of a social milieux, then we run into the familiar, impossible problem. We cannot tell in which group of scientists, G_i or G'_i, science might flourish, or flourish more, from the point of view of the values it has adopted, unless we know the social, political, and economic structure (the social milieux) in which the group exists, and we cannot tell whether the social, political, and economic structure (the social milieux) in which the group exists is vital in order for science to flourish unless we know the values the members of the group hold. Let us call this the Kuhnian Circle.

Second, suppose three kinds of societies: capitalistic, socialistic, and Marxist. Consider the pure version of each of these three societies, and suppose there exists a group of scientists in each of these societies. Assume also the group of scientists in each society to be neither too small nor skewed by inordinate historical circumstances, and assume the three groups of scientists to function along the five values. Each of these is a group of scientists in which "the individual differences distribute on some normal curve," and the scientists do not share atypical sets of values. But, surely, *nothing* follows about how good each group is, or how far removed each is from, say, G'_I. It is difficult to see how Kuhn could make this much-needed claim unless he were to suppose that values, and how they ought to be used, can be defined independently of the external environment, and the success of science likewise gauged. *Only* if we have such a claim would it make sense for Kuhn to speak in terms of inimical values that deter science, and in terms of catalytic values that enhance the growth of science. But this would leave Kuhn's sociological thesis in shambles.

Third, I have left alone the dynamic problem of group rationality and focused solely on the static problem. The static problem of group rationality is to determine whether a group of scientists, at a given time, is structured rationally; the dynamic problem of group rationality is to determine whether a group of scientists has evolved to a rational structure or to a more rational structure.[46] With respect to the latter problem, Kuhn advisedly warns us that here we must exercise "great caution" because, as he says, changes "in the environment where science is practiced can also have fruitful effects on research" as well as deleterious ones. Thus, in terms of our schematic example, one can easily foresee how a group of scientists in one society (producing deleterious values) can be transformed as a result of historical transitions that produce a set of values more conducive to the growth of science. G'_{I-1} might experience just that historical shock that brings it to the top, and G'_I might fall into second or third place. Any solution to the dynamic problem of group rationality, of course, must presuppose a solution to the static problem of group rationality.

Fourth, if the Kuhnian answer, remarkably, is that the groups of scientists are *all* equally good, then we may safely dismiss even the need for any reductive analysis. For there is nothing interesting to learn. We know the lesson antecedent to any further historical investigation: Every kind of society – fourth-century medieval France, nineteenth-century Weimar, or twenty-first-century America – is just as good as any other for producing Kuhnian science. Without further argument, I shall assume this claim to be patently false. Nor, surely, is it the answer Kuhn would offer. What, then? Kuhnians must, at least, outline a social theory that would be of great benefit to science, as *they* see science, and once we have this theory, we shall then be able to distinguish between scientists who act irrationally and those who do not. Our means for distinguishing would be that normative social theory or, shall we say,

[46] This, of course, is a distinction of cardinal importance outlined earlier; see Chapter 1, this volume, 9–10.

a theory of group rationality (Scientific Utopia) underpinned by a powerful vision of a good and a just society (Social Utopia).

Suppose the rationality of a group of scientists is defined in terms of methodology, what Kuhn calls the rules of the philosophers. In the ultimate analysis, it is this normative structure, not the social, political, and economic structure of the society in which the science is embedded, that has a profound impact on science. If its methods, or rules, are effective, science will flourish; if not, it will decline. The five key concepts are linked to rules of the method and the success or failure of science is linked to rules. Imagine a group of scientists that had the best available method and was founded on that method. Let us call this *the ideal normative group*.

The notion of the ideal normative group could be an intriguing notion from the point of view of Kuhn. For there might be not one ideal, but several ideals (Scientific Utopias), even for a single group at a single time. Thus, if the five basic values of accuracy, consistency, scope, simplicity, and fruitfulness are nonranked primary values, then a certain group might be rational if *a*% of the scientists in that group accept a certain theory, *T*, on the ground of accuracy, *b*% accept it on the ground of consistency, *c*% accept it on the ground of scope, *d*% accept it on the ground of simplicity, and *e*% accept it on the ground of fruitfulness. Arguably, then, the same result (the acceptance of the same theory, *T*) might also be produced if *e*%, *d*%, *c*%, *b*%, and *a*% of the scientists in the group accepted it on the grounds of accuracy, consistency, scope, simplicity, and fruitfulness, respectively. And, of course, there are several such alternatives producing the same consequence. Is there a Kuhnian argument that indicates preference for one way, or just a handful of ways, of selecting one theory over another? Or is each way equally good? Might this example not also be easily worked into an argument to show that, even for normal cases, while the philosopher's choice by rule might be immediately adapted to Kuhn's choice by value, the converse is hardly obvious?

Case 1: Suppose historians of science discover that *any* society of scientists in the past that was successful held values that were similar to the values of the ideal normative group (making due allowance for minor fluctuations of description and detail owing to contingent, but nonsignificant, historical factors), and that*any* society of scientists in the past that had failed held values that were at odds with the values held by the ideal normative group. Furthermore, the historians find that the degree of success or failure of a group of scientists is neatly correlated with how near or far its structure (social milieux) was to the structure of the ideal normative group. Let us call this *the Invariance Hypothesis.*

Two things would follow. First, on this hypothesis, success or failure of science would be relatively much easier to explain. Second, such a discovery would result in undermining the claim that external environment determines or shapes values: For it can be shown, if the hypothesis is true, that relatively similar values guided the work of successful scientists, at least in theory selection or theory choice, in vastly dissimilar social milieux. For instance, it might be historically discovered that scientists in societies with quite different class structures accepted Einstein's special or general theory of relativity for roughly the same reasons.[47] One might find that over several, very differently structured societies of physicists, the normal distribution captured the distribution of the differences among scientists with respect to their values and judgments. Thus, one might find that (making due allowance for statistically insignificant deviations) in each group of physicists, x% of physicists accepted the theory of general relativity on the basis of accuracy, y% on the basis of consistency, z% on the basis of scope, and so on. Such a finding would confirm the invariance hypothesis.[48]

Case 2: Of course, it is unlikely that there would be such a perfect correspondence; but something that tends in that direction would suffice. Such an historical task would throw considerable light on the interesting philosophical task of (*a*) discovering new values neither on Kuhn's list nor on the lists of the traditional philosophers of science, or (*b*) showing how the allegedly new concepts are linked to the old ones.

[47] Between 1905 and 1915, hardly any new empirical evidence was recorded in favor of Einstein's special theory of relativity. Yet the probability of success of the rival theories of Lorentz and Abraham "fell in the estimates of most of the members of the European physics community," and that of Einstein's theory correspondingly increased. Likewise, when Einstein's general theory of relativity accounted for the exact amount of the anomalous advance of the perihelion of Mercury, the rival hypothesis "dropped dramatically in the estimates of most of the physics community." Earman, "Carnap, Kuhn and the Philosophy of Scientific Methodology," 25–6.

[48] Such groups of scientists might unwittingly tell us how the so-called primary values of evaluation come to be ranked. For instance, one might find that fruitfulness is always ranked highest by a vast majority of scientists, while simplicity is always ranked last.

The recent efforts to give empirical accounts of the scientific practitioners and their science can make philosophical sense only in light of the foregoing. For if the traditional views of philosophers of science in the field of methodology are only so much armchair speculation, then the new studies must not be merely anthropological reports.

Case 3: Here is an extreme case. We might find that societies in the past and present proliferate a scheme of values that have no apparent or real connection with the values of the ideal normative group, but are nonetheless successful. I do not know how we might then explain the success of science, or see any profit in studying the successful societies of scientists. For one thing, it is unlikely to instruct *us* on how *we* might conduct *our* science, arrange our society of scientists, or learn from the successes and mistakes of the past. For all we know, our scheme of values may be just as unique and just as successful. This scenario is quite unlikely to be true. Let us call this *the Grand Coincidence Hypothesis.*

In sum: If the Grand Coincidence Hypothesis is false, Kuhn might at least indicate *how* he might explain the success of science; on the other hand, if the Invariance Hypothesis is true, how might he distinguish his philosophical approach from the more traditional one?

IV. Negotiations in the Scientists' Original Position

"Negotiation" – as we know – is a term of art among sociologists of knowledge. Kuhn is advisedly circumspect regarding the dogmas of this discipline, especially the dogma that when scientists claim something to be knowledge, that claim is established as a result of deliberations – negotiations – between contending or cooperating parties in a given domain of science; and these negotiations, under the cover of reason, theory, observations, results of experiments, evidence, and the like, are nothing other than expressions of the brute power and interest of the scientists. Nature does not enter into this picture; only power, authority, and interests do. This, says Kuhn, is an "example of deconstruction gone mad."[49] Surely, we all follow Kuhn's lead in condemning this. And yet, Kuhn thinks, it represents a "philosophical challenge."[50]

[49] Kuhn, "The Trouble with the Historical Philosophy of Science," 9.
[50] Ibid.

In the ensuing, I shall explain that challenge, Kuhn's solution to it, my animadversions against that solution, and, finally, invoking John Rawls's idea of the Original Position and some ideas of Bernard Williams, both already encountered,[51] outline what, I think, might be an interesting task for a Kuhnian – indeed, even for the rest of us. All this will be addressed in a brief, exploratory manner to indicate, in particular, why this issue of negotiation is important in the context of group rationality and what the present version of the subjectivist view will have to do to safeguard itself against the charge that the boundary line between it and the skeptic's view is quite difficult to discern.

Let us begin with the traditional understanding of the goal or aim of science, namely, *truth* and *verisimilitude*. The aim of science is to discover particular facts and laws of the universe, and if not true ones, then approximately true ones. Beliefs that are true are claimed to be about, and to correspond to, an external, mind- and culture-independent world.[52] But the notions of truth and verisimilitude are quite obscure, says Kuhn, and they need to be replaced. Beliefs must be evaluated, nonetheless, and that evaluation is generally indirect, since there is rarely an opportunity to confront a single, solitary belief with the mind-independent external world; rather, one places that belief in a network of beliefs and using secondary criteria – such as, accuracy, consistency, breadth of applicability, and simplicity[53] – one evaluates that belief network, and indirectly the solitary belief in it. With respect to the selected beliefs, Kuhn avers, they are simply "the better or best of the bodies of beliefs actually present to the evaluators at the time their judgments are reached."[54] And if not that, asks Kuhn, "[w]hat else is scientific knowledge and what else would you expect scientific

[51] See Chapters 1 and 2 of this volume. In the final chapter, I shall refer to this idea of the scientists' original position as the Council of Scientists; for now, I pay homage to Rawls.

[52] "Within the main formulation of the previous tradition in philosophy of science, beliefs were to be evaluated for their truth or for their probability of being true, where truth meant something like corresponding to the real, mind-independent external world." Kuhn, "The Trouble with the Historical Philosophy of Science," 12.

[53] Ibid., 13. Occasionally other things are mentioned as well: beauty, normativeness, and generality; see Thomas Kuhn, "Rationality and Theory Choice," in his *The Road since Structure*, 214.

[54] Kuhn, "The Trouble with the Historical Philosophy of Science," 18.

practices characterized by these evaluative tools to produce?"[55] (Might one simply answer, "Truth"?)

The philosophical challenge, then, is this. Can one present a philosophy of science that, on the one hand, does not peddle metaphysically suspect concepts, like truth and verisimilitude, and, on the other hand, does not fall prey to the strong program in sociology of science that sees science as nothing but the net result of transactions or negotiations between scientists who are nothing but power and interest brokers? Kuhn claims to have met that challenge. "These traditional criteria of evaluation," – accuracy, consistency, simplicity, and so on – Kuhn admonishes, "have been scrutinized also by the microsociologists who ask, not unreasonably, how, in the circumstances, they can be viewed as more than window dressing. But look what happens to these same criteria – upon which I cannot much improve – when applied to comparative evaluation, to change of belief rather than directly to belief itself."[56] Kuhn claims that one should simply dispense with *knowledge*, understood in terms of truth and verisimilitude, and rely instead upon the established secondary criteria in order to make comparative determinations, and claim no more. Then, since one is talking about *change* of belief rather than just belief, and the metaphysical problems arise only in static cases, not in dynamic ones, comparisons are relatively unproblematic and philosophically respectable. Secondary criteria would enable us to establish something like the following: $paradigm_1$ is *more* accurate, consistent, simpl*er*, has wid*er* applicability than $paradigm_2$.[57] Since that is all we need in science, we need invoke neither the metaphysically problematic notions of truth and verisimilitude nor yield to the strong program.

Once we dispense with otiose problems in metaphysics, and the strong program, the road to developing an evolutionary analogy is clear. "Scientific development is like Darwinian evolution, a process driven from behind rather than pulled toward some fixed goal to which it grows ever closer."[58] To continue with the Darwinian metaphor,

[55] Ibid., 17.
[56] Ibid., 13.
[57] Ibid., 13–14.
[58] Ibid., 14. Even to a casual reader of *The Structure of Scientific Revolutions* this will hardly occasion surprise; see especially 170–3. For reconfirmation, see Kuhn, "A Discussion with Thomas S. Kuhn,' in his *The Road since Structure*, 321–2.

science should not be seen as a single monolithic entity, but rather as sprouting a "complex but unsystematic structure of distinct specialties of species, each responsible for a different domain of phenomenon."[59] Any ground between these disciplines or species is "empty space."[60] Presumably, then, if $paradigm_1$ is *more* accurate, consistent, simpl*er*, has wid*er* applicability than $paradigm_2$, we ought not to say that $paradigm_2$ is true or that $paradigm_2$ has a greater degree of verisimilitude than $paradigm_1$. That would be to engage in indefensible metaphysical talk; and we are done with that. We need only say that $paradigm_1$ is "better or best," with respect to the aforementioned values, "of the bodies of beliefs actually present to the evaluators at the time their judgments are reached."

The metaphysical problem, I aver, is not so easily dispensed with. First, nature plays a significant role in Kuhn's view, since his complaint against sociologists of knowledge was precisely that they have ignored it; and it was Kuhn who said: "The decision to reject one paradigm is always simultaneously the decision to accept another, and the judgment leading to that decision involves the comparison of both paradigms with nature *and* with each other."[61] Even in the context of change of belief, and consonant with his last-cited advice, Kuhn cannot avoid the following three steps: (*1*) Compare $paradigm_1$ to Nature; (*2*) compare $paradigm_2$ to Nature; (*3*) after noting the two results, compare $paradigm_1$ to $paradigm_2$ to determine which paradigm is more accurate, consistent, simpl*er*, has wid*er* applicability than the other. The first two steps involve us in just the kind of moves Kuhn prohibits, because one surely cannot compare two paradigms – say, with respect to their puzzle-solving efficiency – without first ascertaining how each relates to nature.[62]

[59] Kuhn, "The Trouble with the Historical Philosophy of Science," 18.
[60] Ibid., 20.
[61] Kuhn, *The Structure of Scientific Revolutions*, 77.
[62] "Now suppose –. . . that the scientist's aim in selecting theories is to maximize efficiency in what I have elsewhere called 'puzzle solving.' Theories are, on this view, to be evaluated in terms of such considerations as their effectiveness in matching predictions with the results of experiment and observation. Both the number of matches and the closeness of fit then count in favor of any theory under scrutiny." Kuhn, "Rationality and Theory Choice," in his *The Road since Structure*, 209. Elsewhere, Kuhn speaks of "maximizing the precision"; see Kuhn, "Afterwords," 338. Kuhn's thesis then amounts to: If a theory, *T*, has maximized efficiency in a way *T'* has not, has more matches with the results of experiment and observation, and therefore a

Second, treating science as if it were not a monolithic enterprise and splitting it into disciplines as one would distinguish biological species does not help either.[63] For even if science were splintered into complex, loosely connected species – disciplines – each paradigm in a discipline would be *about* a certain bit or piece of the phenomenon that constitutes the domain of that species or discipline, and we would have to ask, with respect to each such paradigm, whether it adequately corresponds to – fits, captures, reflects, represents, portrays, depicts, mirrors, or whatever – that bit or piece of the phenomenon. This would return us to precisely the same problem we were trying to avoid before.[64]

Third, one could take Kuhn's secondary criteria to be essentially epistemic criteria to determine the objective, nonepistemic status of a paradigm. A set of methodological norms, reflecting certain values, may lead to the choice of one theory as holding out promise of being nearer to the truth or a better puzzle solver, thus escaping the charge – leveled by Carl Gustav Hempel (1905–1997) – of being "near-trivial."[65] But that is evidently no help to Kuhn, since he will have none of it.[66] Consequently, what we can say about a paradigm that satisfies certain values or methodological norms is not that it is nearer to the truth or will be a better puzzle solver in the long run; we can applaud it only for satisfying certain values or norms. And that *is* trivial.

Let me finish this sketch. Since this is not a treatise on truth, or metaphysics, let us simply concede to Kuhn, for the sake of argument, that the aim of the scientists is neither truth nor verisimilitude, but to

greater closeness of fit, nonetheless we cannot speak of *T* as having a greater degree of verisimilitude than *T'*. Again, one can speak of *T* matching results of experiment and observation, or having closeness of fit, but not correspondence. Yet, Kuhn wants the notion of truth; see my note 64 to this chapter.

[63] Kuhn, "The Trouble with the Historical Philosophy of Science," especially 16–17, 19–20.

[64] ". . . if the notion of truth has a role to play in scientific development, which I shall elsewhere argue that it does, then truth cannot be anything quite like correspondence to reality. I am not suggesting, let me emphasize, that there is a reality which science fails to get at. My point is rather that no sense can be made of the notion of reality as it has ordinarily functioned in philosophy of science." Kuhn, ibid., 14.

[65] See Carl Hempel, "Valuation and Objectivity in Science" and "Scientific Rationality: Analytic vs. Pragmatic Perspectives." For Kuhn's response to Hempel, see Thomas Kuhn, "Rationality and Theory Choice," 208–15.

[66] Kuhn, "Rationality and Theory Choice," especially 213–14.

find a paradigm that best satisfies the secondary criteria at hand. This, let us say, is the goal of science.[67]

Second, granting the goal of science, the talk of negotiations is recast in an entirely different light.[68] Just how can these negotiations take place in the Kuhnian scheme? This is, without doubt, the single most important question for this version of the subjectivist view – in fact, duly adapted, for any view. Let me explain by way of an analogy. In *A Theory of Justice*, John Rawls places moral agents in the Original Position, casting a veil of ignorance on them. As a consequence, these moral agents are deprived of any knowledge that is "arbitrary from a moral point of view."[69] The moral agents know nothing about their sex, wealth, income, social status, class position, natural assets and abilities, intelligence and strength, rational plan of life, and so on. And in this

[67] Not that the issue of truth or verisimilitude discussed thus far is only of metaphysical interest and has no practical import. Asked to testify in an Arizona trial about creationism, Kuhn declined. His reason: "Look, that one I declined for I think an excellent reason. [The people who approached me were resisting the creationists. I was sympathetic, but] I didn't think there was a chance in the world.... I mean I was being used by the creationists, for God's sake! At least to some extent. And I didn't think there was any way in the world in which somebody who didn't quite believe in Truth, and getting closer and closer to it, and who thought that the essence of the demarcation of science was puzzle solving, was going to be able to make the point. And I thought I would do more harm than good, and that's what I told them." Kuhn, "A Discussion with Thomas S. Kuhn,' 321–2.

A few comments are in order: (*A*) Since only science satisfies the methodological values, and Kuhn claims to know no other value, Kuhn could have claimed that creationism is simply not science. See, especially, Kuhn, "Rationality and Theory Choice," 213–14, and Kuhn, "Afterwords," 338. (*B*) Kuhn could have identified truth (notice the absence of capital 'T') as anything that satisfies the methodological values better than any other theory, and then proceeded to show why scientific creationism, failing to satisfy the Kuhnian methodological strictures, is simply not true. (*C*) If a methodology cannot account for why creationism is not science, what – in a philosophical court – counts as a defect in method? (*D*) One should conclude that Kuhn simply dispenses with the notion of truth altogether, and not that he uses it in a radical, nontraditional way. Alas, for the road not taken: see Kuhn, "A Discussion with Thomas S. Kuhn," 278.

[68] My use of "negotiation" is in line with Kuhn's use of it. The "term 'negotiation,'" says Kuhn, "seems to me just right, except that, when I say 'letting nature in,' it's clear that that's an aspect of it to which the term 'negotiation' applies only metaphorically, whereas it's fairly literal in the other cases." "A Discussion with Thomas S. Kuhn," 317. One can straightforwardly conclude, what is crucial for what is to follow, that it is possible for scientists literally to negotiate among themselves about how they should structure their society in order to negotiate – metaphorically speaking, now – with Nature.

[69] Rawls, *A Theory of Justice*, 72.

position, the moral agents are supposed to reason – negotiate – among themselves. Rawls has them reason themselves into a unanimous agreement on two fundamental Rawlsian principles of justice, namely, the principle of equal liberty and the difference principle; and these are the principles in accordance with which the moral agents would agree to run their society.

Suppose, now, that scientists are placed in a Scientists' Original Position in order for them to negotiate by what principles they would structure their society of scientists. In this position, scientists would be listening to one another. (This would answer the Williams question "To whom is it addressed?" or "Who is supposed to be listening?") If, like their moral counterparts, the scientists are to be placed behind a veil of ignorance, then the first task would be to specify what kind of knowledge scientists will be allowed in their Scientists' Original Position, and what kind will be forbidden. They might, for example, be allowed knowledge of the history of science thus far; the economic, political, and cultural structure of their society; what technology is available. But the scientists would have no knowledge of their own scientific knowledge or disposition; their own vested interest in science; their own place in the social, political, and economic structure of their society; the cultural values they espouse, and so forth. This is incredibly crude and is meant only to hint at the problem of how to construct the Scientists' Original Position. What the scientists are allowed to know will determine what *types of reasons* they can use in their negotiations.[70] It has one significant advantage, too: It undercuts the dogma of the sociologists of science at its root because being ignorant of their place in society, scientists cannot be interest or power brokers.

[70] "In this dispute, the challenge was clear. Any success for the Darwinian scheme would require renegotiating – often with bitter controversy – the lines to be drawn between cultural domains. Science was not yet vested with the authority that would come with the modern era. Its practitioners were exerting themselves to create professional communities, struggling to receive due acknowledgment of their expertise and their right to choose and investigate issues in their own manner." Browne, *Charles Darwin: The Power of Place*, 124. Browne is here depicting the nineteenth-century battle over Darwinism; in particular, the battle – renegotiation – over the proper province or domain of science and religion, "over the right to explain origins." My contention is that, in deciding what should be the structure of their society, scientists must deploy reason that is *independent* of the specialization, domain, or profession that is a contending party. If not, the argument will be circular (as Kuhn feared) and cannot be rationally settled.

The second task would be to show *how* the scientists would converge upon the Kuhnian goal of science. This is no mean task: Minimally, a Kuhnian will have to show what characteristics must be accorded to the scientists, and connected to those characteristics, what reasons will be available to them, from which an agreement over the Kuhnian goal of science would follow. (When this task is accomplished, an answer to the Williams question "Why are they supposed to be listening?" will be a natural fallout. They are supposed to be listening because their achieving the Kuhnian goal of science depends on it.)

Let us carefully distinguish between two types of reasons. One type of reasons enables scientists to determine which theory or paradigm to prefer; these reasons are in terms of the values – such as accuracy, consistency, breadth of applicability, and simplicity – already referred to. Let us call them *scientific reasons.* These reasons enable them to select theories or paradigms, but not what structure to adopt for their society of scientists, since that structure is not a scientific theory. Consequently, there must be another type of reasons in terms of which scientists are able to negotiate and evaluate the worth of a particular structure of a society of scientists. Let us call the latter type of reasons *meta-scientific, methodological, or general reasons.* My aim is not to show the irrelevance of the former in selecting a rational structure – far from it. It is rather to show that while it may significantly aid in the cause, the former type cannot by itself perform the task of structure selection.

The significance of constructing a viable Scientists' Original Position is this: (*A*) If scientists are allowed to know everything in the Scientists' Original Position, including, for example, the paradigms they are respectively wedded to, and the epistemic values they possess in consequence,[71] then this version of the subjectivist view will have to explain just how the scientists will negotiate among themselves to arrive at an unanimously agreed-upon view; indeed, it will have to explain just how convergence on a scheme of group rationality is even possible, let alone that it will occur. This is a vital task for this subjectivist view, because if there is no agreement – a likely prospect, given that scientists know everything concerning themselves in the Scientists' Original Position – then it will be hard put to distinguish itself from the skeptic's view. Supposing agreement to be possible, it must

[71] Kuhn, *The Structure of Scientific Revolutions*, 41.

further show how and why it will converge upon (as Kuhn insists) a group structure that normally allows only a single paradigm.

(*B*) If in the Scientists' Original Position scientists are allowed knowledge about the paradigm in which they work, and about the scientific values they hold, and if furthermore they are agreed upon these values, but know none of the contingent, historical interests and values they possess, or the social, political, and economic structure of their society and their place in it, then this version of the subjectivist view has to explain what connection, if any, it can retain to sociology of knowledge, or, alternatively, in what way its claim to be radically different from the traditional philosophy of science can be sustained.

(*C*) If in the Scientists' Original Position scientists are allowed the knowledge they are allowed in (*B*), and if they converge not only upon the values, but also upon what weights should be assigned to each of the shared values – that is, if the group is essentially manned by a single method – then this version of the subjectivist view, in which scientists do not face "a choice between incompatible modes of community life,"[72] would have to explain how in a time of crisis or pre-paradigm stage, the scientists can possibly align themselves to different theories, "describe and interpret [the phenomena] in different ways."[73]

In my view, (*A*), (*B*), and (*C*) *make it crystal clear not only that, but why, the twin tasks of specifying what constitutes individual rationality and how it is connected to group rationality are profoundly important tasks.* We shall see later, not only in examining the work of Hilary Putnam but also in the final chapter, how this is made vivid.[74]

Third and finally, consider this infamous passage from *The Structure of Scientific Revolutions* that explains the structure of *political* revolutions, ending with the claim that *scientific* revolutions have remarkable affinities with political revolutions:

> At that point the society is divided into competing camps or parties, one seeking to defend the old institutional constellation, the others seeking to institute some new one. And, once that polarization has occurred, *political recourse fails.* Because they differ about the institutional matrix within which political

[72] Ibid., 93–4.
[73] Ibid., 17.
[74] See Chapter 8, this volume, especially sections IV–VI, as well as Chapter 9, especially section II.

change is to be achieved and evaluated, because they acknowledge no supra-institutional framework for the adjudication of revolutionary difference, the parties to a revolutionary conflict must finally resort to the techniques of mass persuasion, often including force. Though revolutions have had a vital role in the evolution of political institutions, that role depends upon their being partially extrapolitical or extrainstitutional events. The remainder of this essay aims to demonstrate that the historical study of paradigm change reveals very similar characteristics in the evolution of the sciences. Like the choice between competing political institutions, that between competing paradigms proves to be a choice between incompatible modes of community life.[75]

Kuhn maintains – *The Structure of Scientific Revolutions* has made it abundantly clear – that normally the society of scientists should be governed by only *one* paradigm.[76] Multiplicity of theories may be touted in pre-paradigm and revolutionary stages, but those stages are exceptions to the rule. The paramount task for a Kuhnian, therefore, is to show how scientists will negotiate themselves into agreeing to structure their society of scientists in a way that only allows a single paradigm; it is entirely a different matter – note – what that paradigm or even what the method should be. What is being asked here is to show what *form* the society of scientists should take, not its *content*.

First, the passage roundly declares that in a political revolution there is an extrapolitical or extra-institutional component; similarly, one might infer, in a scientific revolution there is an extrascientific or extra-institutional component (nonscientific values), too.[77] Second, since the problem of group rationality is defined in terms of method, this version of the subjectivist view must distinguish one method from another, if only in broad terms.[78] Kuhn claims that there are no values (none known to him) other than accuracy, consistency, breadth of applicability, simplicity, and so on. No science can be practiced without

[75] Kuhn, *The Structure of Scientific Revolutions*, 93–4.

[76] For two exceptions, see 148–9, this volume.

[77] This is not a bit surprising, since even with respect to normal science, let alone revolutionary science, Kuhn says, "The pursuit of puzzle solving constantly involves practitioners with questions of politics and power, both within and between the puzzle-solving practices, as well as between them and the surrounding non-scientific culture." Kuhn, "Afterwords," 339.

[78] The fair significance of this point will clearly emerge in the next chapter, section V, wherein is presented the new problem of demarcation, namely, the problem of distinguishing viable methods (not science) from nonviable ones.

these; to abandon these values is to cease practicing science. As he says, the epistemic values "are themselves necessarily permanent, for abandoning them would be abandoning science together with the knowledge scientific development brings."[79] But each scientist can accord entirely different weights to these values,[80] from whence flow intense disagreements among scientists, especially in times of crisis and revolution.[81]

One can, then, easily see that it matters how Kuhn defines method. On the one hand, if the core values – accuracy, consistency, breadth of applicability, simplicity, and such – constitute a *single* scientific method, then at most there are merely variations on that method, variations produced by letting scientists give different weights to different values. There is no other option but for the society of scientists to be structured along the lines of a single best method. On the other hand, if giving different weights to different values essentially produces a different method, then it is possible for the scientists, in the Kuhnian framework, to agree to structuring their society along the lines of *multiple* methods. This will lead to an inexplicable result: How can scientists, according different weights to different values, possibly triangulate on a single paradigm?

But if (*a*) nonscientific values are already operational, as the passage declares;[82] (*b*) in the solution of a scientific problem, the scientific values "need not all be satisfied at once";[83] (*c*) each scientist holding these values accords them different weights;[84] and finally, (*d*) these values define what it is to do science,[85] then – leaving aside entirely the question of how the Scientists' Original Position is to be set up – the *far* more interesting question is this: Given (*a*) through (*d*), how can the scientists possibly come to agree on how their scientific society should

[79] Kuhn, "Afterwords," 338.

[80] "Nor does it even imply that the reasons for choice are different from those usually listed by philosophers of science: accuracy, simplicity, fruitfulness, and the like. What it should suggest, however, is that such reasons function as values and that they can thus be differently applied, individually and collectively, by men who concur in honoring them." Kuhn, *The Structure of Scientific Revolutions*, 199.

[81] See Section II of this chapter.

[82] See also section I of this chapter and Figure 6.1.

[83] Kuhn, "Afterwords," 338.

[84] Kuhn, *The Structure of Scientific Revolutions*, 199.

[85] Kuhn, "Afterwords," 338.

be structured? In short, how would this version of the subjectivist view solve the problem of group rationality any differently than the skeptic solves it?

Or one might look at this interesting question from the standpoint of the Williams problem: A theory of group rationality is not offered in a vacuum. In attempting to justify itself, it is attempting to do so against a certain background; in the foregoing, I have tried to show that this background must be the one provided by the skeptic's view. If this second version of the subjectivist view is successful in this task, it will also have provided an answer to the Williams question "Against what?" For recall that Williams argued that in the moral case, a justification for being ethical is arguing against the motivations of an amoralist. By parallel reasoning, a subjectivist attempting to justify his theory of group rationality might be considered to be arguing against the skeptic.

V. The City of Man?

To answer the Williams problem about Social and Scientific Utopias[86] from the present perspective, let us turn to an apparently innocuous chapter from *The Structure of Scientific Revolutions*: Chapter 11, "The Invisibility of Revolutions." So innocuous has it appeared that it is the least commented upon of all the chapters of that astonishing, puzzling book; it is something one might comment upon in passing, but it is hardly ever made the subject of an extended paragraph, let alone a paper.

Revolutions in science are necessary. They are changes of world-views, and momentous though they are, they are invisible. And, says Kuhn, there are excellent reasons for their invisibility. Of these, Kuhn mentions three: textbooks in science, popularizations, and philosophical models. Both popularizations and philosophical models take their cue from the textbooks. By far the most important, therefore, is the textbook tradition. It takes as unquestioned the body of data, problems, and theory that governs current scientific practice, with little or no attention to their historical evolution, and attempts to communicate these to students and professionals alike. "Textbooks thus begin

[86] See Chapter 1 of this volume, 19.

by truncating the scientist's sense of his discipline's history and then proceed to supply a substitute for what they have eliminated."[87]

Here, then, are some untruths, lies, around which the life of a student or professional scientist is built. First, "students and professionals come to feel like participants in a long-standing historical tradition." Yet that tradition has, "in fact, never existed,"[88] that feeling of belonging is anchored in untruth. What function does that feeling of being a participant in a tradition serve? What does identifying with the heroes of a science's past do for the current practitioner? Is that feeling, or identification with past heroes, indispensable for him to function as a scientist? If so, in what way? Could that feeling be nourished in a different way? How about encouraging an individual's sense of belonging to the present society of scientists? Might a closer bond result? If the truth were known, and that feeling disappeared, how would it affect scientific practice? Could not something else – a different feeling – serve as a surrogate?

Second, "the scientists of earlier ages are implicitly represented as having worked upon the same sort of fixed problems and in accordance with the same set of fixed canons." This, too, is an untruth perpetrated "partly by selection and partly by distortion."[89] And this fraud is perpetrated after *each* revolution. So long as this is merely story telling, with no other overt aim, it is harmless enough. But, if so, why are the distortion and fraud not generally identified? Why produce this false sense of continuity? Or does this lie, namely, that this generation is working on problems and with canons as did the scientists of yesteryear, perform a function, like the last one, that does not have a substitute?

Third, "human idiosyncrasy, error, and confusion,"[90] which a more detailed historical narrative would reveal, is ignored. Reason for this untruth: "Why dignify what science's best and most persistent efforts have made it possible to discard?"[91] What is genuinely error or confusion should, indeed, be discarded; what is peddled as confusion, but is not, is a different story. *That* is not dignifying error so much

[87] Kuhn, *The Structure of Scientific Revolutions*, 137.
[88] Ibid., 138.
[89] Ibid.
[90] Ibid.
[91] Ibid.

as ejecting an inconvenient truth. One should, in textbooks and in pedagogy, teach, disseminate, and proclaim the truth. One should say, "Standards have changed, problems have altered, and what is thought to be confusion is not only not so, but to the contrary, history shows that they were very plausible beliefs when placed in proper historical context." Why, then, is it essential to perpetuate the illusion that what is now discarded must have been human idiosyncrasy, error, and confusion?

This culminates, fourth, in a statement that can only be described as utterly enigmatic: "The depreciation of historical fact is deeply, and probably functionally, ingrained in the ideology of the scientific profession, the same profession that places the highest of all values upon factual details of other sorts."[92] One ought not to blame a scientist for being uninterested in the history of science. But given that he is interested in it, how can his otherwise innate drive for details and veracity – his desire to search for fitness – not compel him to seek similar things when it comes to the history of his discipline? Textbook writers prefer to ignore, select, bias the data in favor of whatever it is that they are writing about, say, the notion of the chemical element. Why is there not any preference, given the innate disposition of the scientists, to sell, or to receive, a more accurate version? Why are textbooks in the history of mathematics less susceptible to such foibles? Why could scientific textbooks not be written, like textbooks in mathematics, and just as successfully, without the least bit of consideration for the past, without mention of heroes, dates, and historical trivia? Would the student or the professional learn less from such a textbook? Why would such a textbook writer not be less guilty of historical anachronism?

Evidently, history is required, is indispensable, even if it is history that is distorted by the temptation to write Whiggish history – a temptation, warns Kuhn, that visits scientists more often than it does members of other professions.[93] Every schoolboy learns to ask the reason behind Rousseau's lie. In Rousseau, the populace or the masses, not much inclined to be legislators, are not educated enough, nor cultured enough, nor do they have enough insight or foresight, to understand

[92] Ibid.
[93] Ibid.

the wisdom of the laws proposed by the Legislator. Hence, the lie.[94] But at least the lie is intended for their good. What is the point of the lies here? First, the scientists are educated enough to understand why science is not cumulative, linear; cultured enough to understand how and why aims, problems, and canons of inquiry change from age to age; and insightful enough so that, if a curious scientist were to raise questions about this version of relativism, they could be confronted with historical data and problems that were sufficiently appealing to their sense of detail and veracity.

Second, *what* is the good intended in this case? How do the scientists benefit from the lies that are propagated in their midst, a good or benefit they would not receive if the truth were known? Why could scientists not be raised in the belief, by proper and more truthful instruction in the history of their discipline (as Kuhn views it), that there is little or no genuine continuity with the past, and that what they do now is not necessarily progress over the past achievements in science? Having learned that lesson, how and why would they become of diminished capacity to do science? If not, why perpetuate those lies? Until those questions are plausibly answered, the fraud, the distortion, and the prejudicial selection are hard to comprehend, let alone our admitting them, with our blessing, as forces that should shape a society of scientists.

Let us shelve this point for now, and return to the issues of the last section. Therein we asked, "Which social structure, or what kind of society at large (Social Utopia), would be most beneficial for Kuhnian science?" The reason why this philosophical task would be of value – perhaps even of great importance – is because such a theory of society in which Kuhnian science should be embedded will enable an individual scientist to determine for himself, or herself, what is his or her worth in the Kuhnian scientific enterprise; what role and value he or she has in the social scheme; and, lastly, given that scheme, whether he or she would be a willing participant. Not insignificantly, then, surely, when scientists are negotiating (see section IV), these issues will be in the forefront of their concerns. Kuhn made significant the investigations into the history of science, focused our attention on the dreary

[94] Jean-Jacques Rousseau, *On the Social Contract*, 38–41.

side of science, and claimed that that is supposed to benefit science as a whole. Perhaps this is so, perhaps not. Yet until we are told what the inspiring vision (Scientific Utopia) is that can enlist the willing participation of all the members of science, we shall never be able to determine, in the final analysis, whether that allegiance is something that a scientist should give or should refrain from giving.

Kitcher has gone a step further, and his view of group rationality alarms me even more: Not only must we do this, but we had better, says he, take into account the baser motives of man, and of scientists in particular, and tap these motives so that we might fashion a more rational society. He makes virtue of our vice, like a latter-day author of *The Fable of the Bees.* His argument is not that a philosophy of science constructed on ideal assumptions is all very well, but that it is uninteresting because it does not reflect what we are likely to encounter in the real world. His argument is rather that we should be *thankful* that such an ideal is not to be discovered there: For if scientists were ideal – epistemically pure, as he puts it – they would have botched the collective enterprise. What we therefore need to do is to capitalize on the shabby moral traits and characteristics scientists have and show how these can be harnessed to the success of the scientific group. A theory of group rationality need not do more than that; and if Kitcher is right, then indeed it cannot do more.

I do not know how baser motives will react when given that degree of prominence in the pursuit of truth, or what repercussions they will wreak on the society of scientists, and on our society at large. So I am skeptical of such a move. It might be more useful to show how ideal scientific men, a veritable society of Rousseauean scientists, would negotiate: how they might arrange their affairs rationally, what assumptions they would make, what moves prescribe, and thus teach us how we might mimic their ways insofar as our human foibles would allow.

Georg Wilhelm Friedrich Hegel ends his preface to *The Phenomenology of the Mind* with the following:

> For the rest, at the time when the universal nature of spiritual life has become so very much emphasized and strengthened and the mere individual aspect has become, as it should be, correspondingly a matter of indifference, ... the share in the total work of mind that falls to the activity of any particular individual can only be very small. Because this is so, the individual must all the more

forget himself, as in fact the very nature of science implies and requires that he should; and must, moreover, become and do what he can. But all the less must be demanded of him, just as he can expect the less from himself, and may ask the less from himself.[95]

Whatever I have conceived the problem and theory of group rationality to be, it is not meant to render the individual scientist insignificant, to make him subservient to science, or to enslave him to a whole larger and more significant than himself. Hitherto, if we have failed to address the problem of group rationality, our failure must have seriously affected our understanding of what makes for, or creates the conditions for, the well-being of the individual, or at the very least, what aids or assists the individual in his search for knowledge (a search that is intimately tied to his well-being).[96] Correspondingly, my hope has been that if we do succeed in finding an adequate theory of group rationality, it will enable us to design or construct a society of scientists accordingly. The immediate beneficiary of that design or construction would be the individual: the individual alone before God – or, if you prefer, before gods, nature, or nothing. What does not promote his well-being, in the ultimate analysis, has no bearing on what I am doing in this book.

It is fitting, therefore, that the last word should go to an arch anti-Hegelian:

It is Christian heroism – a rarity, to be sure – to venture wholly to become oneself, an individual human being, this specific individual human being, alone before God, alone in this prodigious strenuousness and this prodigious responsibility; but it is not heroism to be taken in by the idea of man in the abstract or to play the wonder game with world history. . . . The law for the development of the self with respect to knowing, insofar as it is the case that the self becomes itself, is that the increase of knowledge corresponds to the increase of self-knowledge, that the more the self knows, the more it knows

[95] Cited in Søren Kierkegaard, *The Sickness unto Death: A Christian Psychological Exposition For Upbuilding and Awakening*, 173–4.

[96] Please note: Unlike the skeptic, I do not make the scientist's welfare, flourishing, or well-being the basis of structuring the society of scientists; see, for example, 75–6, 95, and 103. On the contrary, I anchor the scientist's well-being in his science. Nor am I being guilty of making the false claim that his science is a sufficient condition of his well-being. Surely, there are several other significant factors beyond science that are necessary as well.

itself. If this does not happen, the more knowledge increases, the more it becomes a kind of inhuman knowledge, in the obtaining of which a person's self is squandered, much the way men were squandered on building pyramids, or the way men in Russian brass bands are squandered on playing just one note, no more, no less.[97]

Thus, Søren Kierkegaard.[98]

[97] Kierkegaard, *The Sickness unto Death,* 5 and 31.

[98] A discerning reader might be puzzled. What justifies, he or she might well ask, my appeal to an arch subjectivist in philosophy of religion while I have been so relentlessly critical, for two chapters in a row, of subjectivists in philosophy of science? My answer is in two parts: First, I am sympathetic to the value Kierkegaard places on the individual human over impersonal things like the forces of world history. And that is the point that is being sold here. Second, Kierkegaard's subjectivism in religion, on the other hand, I have scrutinized, and criticized, in *The Toils of Understanding: An Essay on "The Present Age."*

7

The Objectivist View

Sharply contrasting with the subjectivist views of the last two chapters is the objectivist view. What, then, is the objectivist view? That view suggests that there is a mind-independent reality, and that not only scientific theories but philosophical ones as well are deemed adequate only if they capture, reflect, mirror, correspond to, depict – or whatever – that reality. The view has descended, not without transformation, from Plato. To many it has seemed an exceedingly plausible view, and it is taken to have solved some deep issues in philosophy of science. In methodology, there are two objectivist views that can be fairly regarded as classical: the inductivist view and the deductivist view. The first was given birth by Francis Bacon; the second, three centuries later, by Karl Popper. My interest is not in the differences, the details, or the historical development of these views. It is rather in what these views *would* have said, *if* they had considered the problem of group rationality. My aim is to show that, given their respective traditions, they would have offered inadequate theories of group rationality. I shall argue primarily against the second deductivist view, but my arguments also apply to the first, because despite their dissimilarities, they have a lot in common.

To the question "Under what conditions is a scientific society rational?" the objectivist view would reply, "It is rational when the society is governed by the single best available method." In section I, I shall analyze the problem of group rationality from the vantage point of the objectivist view, and conclude that there are two distinct problems. I shall focus on only one of them, and on its solution. In section II, I shall present the path to the best available method; the next section will present the best method itself. In section IV, I shall demonstrate that the objectivist solution of structuring a society along the lines of a single method is inadequate. It is inadequate because it is too risky and because such a structure would fail to let the society of scientists learn from its methodological mistakes. Next, I shall introduce a new problem of demarcation in section V, and offer an illustration in the next section of why this new problem is significant, using an example drawn from the herbalist tradition. In the final section, I shall examine why an attempt to solve the problem of group rationality appears to be a dangerous enterprise, given that such a task appears to have a striking resemblance, at first glance, to the problem of social engineering. The fears, I shall argue, are ill-founded.

I. Two Objectivist Problems

What was once sought was certain knowledge. When this was found to be unattainable, a more modest goal was pursued: to find theories with a high degree of probability. As a result, a variety of methods were offered, from Bacon to Bayes, that attempted to determine the probabilities of theories based on total available evidence. The problem of group rationality, therefore, can be formulated with this fundamental aim in view.

What if this aim is denied? What if it is argued that we can never reach the truth; or that if we reach the truth, we can never know it; or that we can never even have probable knowledge, for reasons that are as clear and cogent as the ones given by David Hume? What if the best we can do, as Popper asserted, is to falsify our theories, not to verify them? What, then, if the falsification of our theories were the only sensible goal for scientists to pursue? How would such a different goal lead us to revise or restate the problem? Or should we now simply abandon the problem of group rationality altogether on the ground that it is based on false assumptions?

Not quite. Let us begin with our formulation of the problem of group rationality:[1]

[1] To ensure readability and avoid unnecessary complexity, I have presented the problem duly truncated. The full statement occurs earlier in Chapter 3, 67.

Engaged in a joint enterprise to pursue a set of shared (partially overlapping?) goals, what ought to be the structure of a society of scientists, in virtue of which that structure would best enable that society of scientists to reach its goals better than it would under any other structure?

Two instances of that problem easily follow. I shall call the first instance Bacon's problem, and the second instance Popper's problem. The first instance:

Bacon's Problem: Engaged in a joint enterprise to pursue theories that at least have a high degree of probability of being true or verisimilar, what ought to be the structure of a society of scientists, in virtue of which that structure would best enable that society of scientists to reach its goals better than it would under any other structure?

The second instance:

Popper's Problem: Engaged in a joint enterprise to pursue theories that are true or verisimilar, and to eliminate false or unverisimilar theories, what ought to be the structure of a society of scientists, in virtue of which that structure would best enable that society of scientists to reach its goals better than it would under any other structure?

As yet, these two problems do not show that the problem of group rationality should be dissolved. Far from showing that, they call for a solution. Indeed, Popper's methodology of falsificationism can be regarded as a solution to the second problem; more, Popper would assuredly argue that no scientific society founded on any other method – since these would be susceptible to Hume's well-known arguments against induction – could be rational. On the other hand, those offering a solution to Bacon's problem would argue for their theory against Popper's. Their principal defense: Popper himself has not escaped making inductivist moves,[2] and given that inductivist moves must be made, the moves they make are better than Popper's.[3]

Whatever the differences between Bacon's problem and Popper's problem, the two problems, and their respective solutions, share at least three things in common. First, both versions claim that the

[2] Among the vast literature against Popper and John Watkins's "Hume, Carnap and Popper" on this point, Imre Lakatos's "Popper on Demarcation and Induction" and Wesley Salmon's "Rational Prediction" are little masterpieces.

[3] Essentially, Lakatos's position in "History of Science and Its Rational Reconstructions."

solution to the problem of group rationality must be hemmed in by the principles of logic. The falsificationist version of the problem, for example, presupposes two-valued logic. It presupposes this because of the enormous importance it places on criticism and critical tradition. Any method that presupposes a higher-valued logic, it says, is too weak and insufficiently critical.[4]

Second, neither version maintains that new alternative methods will not be forthcoming or that the present method will not be replaced in the future, any more than either view claims that new theories of science will not flourish or that the ones flourishing now will not be superseded in the future. Consequently, the solution proposed is conjectural in nature, not one offering certainty.

Third, the common content of their solutions to the problem of group rationality lies in claiming that the society of scientists should be structured on a single best available method. From the vantage point of Popper's problem, the best method must be deductive. Popper would argue for his method. From the vantage point of Bacon's problem, that method must be inductive. For instance, Hilary Putnam,[5] Imre Lakatos, Thomas Kuhn, and John Watkins would each argue for their respective inductivist methods. I shall treat only Popper's deductivist method. But my arguments would serve just as well against the inductivists. My central argument is that the common content of the objectivist view is false.

[4] "Now, what I wish to assert is this. If we want to use logic in a critical context, then we should use a very strong logic, the strongest logic so to speak, which is at our disposal; for we want our criticism to be *severe*. In order that the criticism should be severe we must use the full apparatus; we must use all the guns we have. Every shot is important. It doesn't matter if we are over-critical: if we are, we shall be answered by counter-criticism. Thus, we should (in the empirical science) use the full or classical or two-valued logic. If we do not use it but retreat into the use of some weaker logic (as Reichenbach suggested in connection with quantum theory) – then, I assert, we are not critical enough; it is a sign that something is rotten in the state of Denmark (which in this case is the quantum theory in its Copenhagen interpretation)." Karl Popper, *Objective Knowledge*, 305–6.

[5] I realize that Putnam is not happily placed in the field of methodology created by Popper, Thomas Kuhn, and Imre Lakatos; he marches in other pastures. Yet his early work can be fruitfully developed as if it belonged there. One such effort is my paper "Putnam's Schemata," which is on Putnam's unjustly neglected piece, "The 'Corroboration' of Theories" (*Philosophical Papers*, Volume 2). For Putnam's defense against Popper, see Putnam, "A Critic Replies to His Philosopher," and *Words and Life*, 465–6. Putnam's later views are presented in detail in Chapter 8 of this volume.

II. Toward the Best Available Method

Popper's methodology, which is widely known, has received severe criticism – not all of it fair – but his meta-methodology has suffered from benign neglect.[6] Given my present concerns, I shall attend to Popper's method and to his meta-method only to the extent that they bear upon the problem of group rationality. In *Conjectures and Refutations*, Popper considers four central epistemologies, analyzes them, and rejects them in favor of the method of falsificationism.[7] These epistemologies are epistemological relativism,[8] epistemological pragmatism,[9] epistemological optimism,[10] and epistemological

[6] I have treated Popper's meta-methodology at length in *A Theory of Method*, Chapter 2; here I shall confine myself to those aspects of his meta-method not treated there.

[7] The summaries of the four epistemologies I now present in the following four footnotes are discussed in detail in Karl Popper, *Conjectures and Refutations*, 4–12.

[8] Epistemological relativism is the doctrine that there is no objective truth. The doctrine is generally regarded as hung on the horns of a dilemma: If the doctrine is true, then there is an objective truth after all, and the doctrine is false. If the doctrine is false, one need not accept it.

[9] Epistemological pragmatism confuses truth with usefulness. This doctrine has a pernicious effect in law. If there is no objective truth, or if truth is the same as usefulness, then the judge need not interpret law so as to capture the spirit of the law, or attempt to make a correct interpretation of the law. For lacking objective truth, the judge can no more be right in his interpretation of the law than he can be wrong. Worse yet, where truth or right are synonymous with what the ruling majority believes or ordains, then such an epistemology sanctions the founding of a totalitarian state.

[10] Epistemological optimism had its roots in the Renaissance. This doctrine had a powerful effect on the growth and development of the human mind and institutions. Its basic tenet was: Truth is manifest. This view overrated the powers of human reason and experience. Indeed, so much so that on this view it became necessary to explain how it was that humans could err, given that if reason and experience were free of passions and prejudices, they would uncover truth. It was concluded that minds were rarely free of passions and prejudices. This was one step away from the conspiracy theory of ignorance, which was required to explain the corruption of minds that led to errors in judgment, experience, and reason.

Epistemological optimism will be intolerant. If error and misjudgment are not the natural outcome of an inquiring, unhampered mind, but are a consequence of the actions of the conspirators, then clearly it is the first task of designing a just and safe society to keep the conspirators out. Where virtue, justice, and fairness have lost, or appear to be on the verge of losing, it is time to seek out the culprits, and to hold them at bay, so that the work of the society can proceed apace and at peace. Similarly, in designing a rational society of scientists, due allowance must be made of the possibility that there will be conspirators in the society, holders and disseminators of outdated theories and poor methods. Truth may have lost, or appear to be on the verge of losing, to these conspirators. Then these conspirators, too, will have to be

pessimism.[11] Popper's conclusion is that none of these four methods is adequate (and, hence, none would serve to found the structure of a rational society of scientists). What method is left for the task?

While these epistemologies cast out authorities in knowledge, they only succeeded in replacing them with a different authority because they implicitly asked the same question, "What should be our authority in the search for knowledge?" Accustomed to appealing to authority, these epistemologies asked questions that, once again, demanded an answer in terms of authority. While Baconian epistemology turned to the senses as its authority, the Cartesian epistemology invoked the authority of the intellect.

One rationale for the search for right authority is the hope of discovering certainty. But what if knowledge, and scientific knowledge in particular, can never be certain? Then, says Popper, the old epistemologies "failed to solve the great problem: How can we admit that our knowledge is a human – an all too human – affair, without at the same time implying that it is all individual whim and arbitrariness?"[12]

Since antiquity, political philosophers have asked, "Who should rule?" The question begs for an authoritarian answer. "The best," "the wisest," "the philosopher-king," "the people," "the majority" were considered as candidates. But this traditional question in political philosophy should be replaced by another question, "*How can we organize our political institutions so that bad or incompetent rulers* (whom we should try to avoid having, but whom we might so easily get all the same) *cannot do too much damage?*"[13] Only by pursuing the latter question is there

eliminated by the powers that be, so that the task of science can proceed without hindrance. Recall scientific creationism and Peirce's method of authority.

[11] Epistemological pessimism errs in the opposite direction. It avers that nothing can be known because the mind is too depraved or inadequate an instrument. And it, too, will have totalitarian consequences. If humans are incapable of governing their moral lives, then someone capable has to be placed at the helm. And he, or a select few, must then be vested with the power to dictate to the ill-possessed, handicapped majority what is right and what is wrong, which institutions are permitted and which ones are not, what shall be the roles and rights of various individuals, and so on. Think of Plato or Sidgwick.

[12] Popper, *Conjectures and Refutations*, 16.

[13] Ibid., 25. This view is so implausible that it is not surprising that, referring to what he calls the "ambivalence of social institutions," Popper later in the book (133–4) says that "a social institution may, in certain circumstances, function in a way which strikingly contrasts with its *prima facie* or 'proper' function. ... The ambivalence can

any hope, on Popper's view, that we will discover a reasonable theory of just and effective political institutions.

But as in political philosophy, so in epistemology.

> The question about the sources of our knowledge can be replaced in a similar way. It has always been asked in the spirit of: "What are the best sources of our knowledge – the most reliable ones, those which will not lead us into error, and those to which we can and must turn, in case of doubt, as the last court of appeal?" I propose to assume, instead, that no such ideal sources exist – no more than ideal rulers – and that *all* "sources" are liable to lead us into error at times. And I propose to replace, therefore, the question of the sources of our knowledge by the entirely different question: "*How can we hope to detect and eliminate error?*"[14]

In a slogan: Ask not for authority, ask for ways of eliminating mistakes.

The rationale of the Popper problem of group rationality can now be better understood. Scientists should be engaged in a joint enterprise to pursue theories that are true or verisimilar; they should aim at eliminating false or unverisimilar theories; and they should try to eliminate them more quickly and efficiently than they could under any other alternative structure. Such a structure of the group of scientists should neither overrate the capacities of the scientists nor underrate them; it should be neither too optimistic nor too pessimistic. Presumably, then, Popper's method of falsificationism – the objectivist view – as

undoubtedly be much reduced by carefully constructed institutional checks, but it is impossible to eliminate it completely. The working of institutions, as of fortresses, depend ultimately upon the persons who man them; and the best that can be done by way of institutional control is to give a superior chance to those persons (if there are any) who intend to use the institutions for their 'proper' social purpose."

What, then, would Popper have us do? Should we place checks and balances on institutions so that bad or incompetent rulers cannot do too much damage? Or shall we place checks and balances on institutions so that competent rulers are given a better chance to man the system? Is there *any* argument to indicate that one can create institutions that can do *both*? – namely, prevent bad people from doing too much harm and present an environment in which good people can flourish? If not, then the argument by analogy fails. I conclude: Even if the asymmetry is applicable when dealing with theories, it is difficult to see how it is applicable in the present case – or in the case of the problem of group rationality. Will we be asked to devise a society of scientists such that its structure will prevent poor methods from taking over? or will we be asked to provide the proper conditions to enable good methods to flourish? Once again, could there be an argument for asking us to do both?

14 Ibid., 25.

a provider of structure will claim to be adequate on all these grounds. Let us see if it is.

III. The Best Available Method

There is no statement of the problem of group rationality, let alone of its solution, in the objectivist tradition. Popper has argued that he has a solution to the problem of induction, duly reformulated, while the original problem is susceptible to no solution, in principle; one need only read *A Treatise of Human Nature* to know why. Popper's solution consists in replacing the principle of confirmation or verification with the principle of falsification. His critics were unsatisfied. Part of the dissatisfaction arose not so much from the bald solution as from the subsequent and substantive theory that was offered pertaining to the problem of demarcation, the evaluation of theories so demarcated, the growth of knowledge, the degree of corroboration of theories, the nature of severe tests, the ability to weed out false theories, and so on. It seemed none of Popper's views worked. Or the inductive principle was tacitly invoked.[15]

But consider a different methodological claim. There are three requirements, says Popper, for the growth of knowledge: First, the new theory should be a simple, powerful, and a unifying idea in the manner of the theories of Galileo and Einstein. Second, the new theory should be independently testable. Third and finally, the theory should actually pass some new and stringent test, such as that of Dirac – every elementary particle has an antiparticle – or Yukawa's pertaining to the existence of mesons.[16] The third test is a strong material requirement, for a good deal rides on satisfying it. For instance, failure to satisfy the third requirement, at least occasionally, would mean the impossibility of further progress in science. If none of the new predictions or new effects – such as that under certain circumstances planets would deviate from Kepler's laws, or that light, its zero

[15] Then there were well-known logical problems pertaining to the aim of verisimilitude, comparison of theories either with respect to their empirical content or with respect to their question-answering capacity, and such.

[16] Popper, *Conjectures and Refutations*, 240–8.

mass notwithstanding, would be subject to gravitational force – were verified, then there would be no increase in our knowledge. In one sense, the third requirement is not indispensable. A refuted theory – like the Bohr, Kramers, and Slater theory of 1924, which was quickly refuted by the coincidence experiments of Bothe and Geiger – could increase the stock of problems and unexplained experimental facts, and hence give an impetus to science.

However, the third requirement is indispensable in this way. If all our attempts at verifying the novel predictions failed, we would be left with an "unbroken sequence of refuted theories which would soon leave us bewildered and helpless: we should have no clue about the parts of each of these theories – or of our background knowledge – to which we might, tentatively, attribute the failure of that theory." But, says Popper, "It seems to me quite clear that it is only through these temporary successes of our theories that we can be reasonably successful in attributing our refutations to definite portions of the theoretical maze. For we *are* reasonably successful in this."[17]

So we are. The problem is how we are to identify the troublesome theories in the theoretical maze; in short, what fallible methodological conjectures do we use? Or are we just simply *lucky*[18] in this, and nothing more can be said on the subject? Suppose we had two theories, *T* and *T'*, and wished to conjecture which of them was nearer to the truth; how would we do so? Popper offered "a somewhat unsystematic list of six types of cases in which we should be inclined"[19] to conjecture just that. *T* is nearer to the truth than *T'* if (*1*) *T* makes more precise assertions than *T'*, and these assertions stand up to experimental tests; (*2*) *T* explains more than *T'*; (*3*) *T* explains the facts in greater detail than *T'*; (*4*) *T* has passed tests that *T'* has failed; (*5*) *T* has suggested new experimental tests, neither dreamed of nor devised before *T*, and *T* has passed these tests; and finally, (*6*) *T* has connected what hitherto appeared to be unconnected problems.[20]

[17] Ibid., 243–4.
[18] Ibid., 8.
[19] Ibid., 232.
[20] Popper's third requirement is seriously problematic; for argument in support of this claim, see Husain Sarkar, "Popper's Third Requirement for the Growth of Knowledge."

Popper's theory of method – his meta-methodology – claims that methodological statements are mere conventions, and that competing methodological claims or conventions can be judged by their fruitfulness.[21] One way, I take it, to demonstrate the fruitfulness of one's method is to show that the method can help to avert the possibility of being saddled with an unbroken sequence of refuted theories. We are successful in averting such a possibility, and the method shows how. To be sure, our methods are merely conjectures and not guarantors of success; whether we are successful in meeting the third requirement is not in our hands (especially so if our method is mistaken). It depends on the methodological reality, although on Popper's view both our successes and our failures must remain mysterious.[22]

"If there is a true theory," says Popper,

> of the kind we are looking for, then the method of proposing bold guesses and of trying hard to eliminate the mistaken ones *may* be successful in producing a true theory; and it clearly does not leave everything to chance, as does the method of bold guesses alone, without attempted elimination; or worse, the method of just sticking dogmatically to one guess, to the one theory that first occurred to us; or still worse, the method of having no theory at all.[23]

Worse, and hence less rational, methods do exist. Popper's meta-conjecture, call it meta-conjecture 1, is that his is the best of all available methods.

> I do not attempt to "justify" the method – I am not a justificationist. On the contrary, I have always stressed that it *may* not succeed; and I shall be surprised but happy if anybody suggests a better or more rational method and one whose success is certain or probable.[24]

Consequently, Popper would assuredly argue that the solution to the problem of group rationality is simple and straightforward: The structure of the scientific society should be founded on the methodology of falsificationism, the method of bold conjectures and refutations. This is a risky enterprise, but one that may succeed.

[21] Karl Popper, *The Logic of Scientific Discovery*, Chapter 2, section 11.
[22] For details, see Sarkar, *A Theory of Method*, Chapter 2.
[23] Popper, *Objective Knowledge*, 365.
[24] Ibid., 366.

IV. Is the Single Method Sufficient?

Are all extant methods deficient in the ways Popper has delineated? Here is a reconstruction of Imre Lakatos's method of research programs, shorn of its whiff of inductivism.[25] Imagine Lakatos saying this: "Whether truth or verisimilitude can be defined or not, we cannot reach it; nor would we know that we had reached it, if we did. The heuristic advice of our method is simply a convention, and no matter how successful we are in giving this advice, we cannot claim it to be true or verisimilar. Our task is simply to make mistakes as quickly as possible so that we can eliminate false or unverisimilar theories. Popper's meta-conjecture 'leads us to expect that the method of conjectures and refutations may be helpful in order to answer questions in certain fields';[26] and this meta-conjecture, says Popper, is fallible. I think Popper's meta-conjecture is not only fallible but false. Popper's convention can be logically opposed by a consistent rival convention. Here is mine."

What follows is the reconstruction of the methodology of research programs. Let the unit of methodological appraisal be a scientific research program, and not individual theories. A scientific research program is divided into two parts. One part is called the hard core or the negative heuristic; the other part is called the protective belt or the positive heuristic. When a prediction fails, the failure is to be attributed by methodological convention to some auxiliary theories in the protective belt, and not to a law or laws in the hard core. Over an interval of time, the number, nature, and character of the auxiliary theories in the protective belt will change.

As an example, consider Darwin's theory.[27] The theory of evolution and the principle of natural selection are to be regarded as part of the hard core or the negative heuristic. Among the conjectures or

[25] Imre Lakatos, "Popper on demarcation and induction," 159–67.

[26] Popper, *Objective Knowledge*, 366.

[27] From Lakatos's letter dated January 7, 1972 (personal communication): "I have been for some time trying to interest somebody in studying the development of Darwinian theory as a research programme without success. There is no doubt that it is a research programme; on the other hand it is very doubtful whether it was empirically progressive from Darwin's day until the neo-Darwinians of this century." For some interesting arguments against the epistemology of Lakatos, especially in connection with Darwin's theory as a research program, see Putnam, *Words and Life*, 472–5.

auxiliary hypotheses in the positive heuristic or the protective belt are his theory of pangenesis; the available theories in geology under the title of uniformitarianism; recent results concerning the structure, physiology, and morphology of animals, and so on. There were a host of problems facing Darwin's theory of evolution. For instance, how is the lack of continuity between successive species to be accounted for? How are remnant characteristics and structures to be explained? How did mimicry and camouflage arise? How is the sudden radiation that occurred during the pre-Cambrian to be explained? How is the process of evolution to be squared with geological time?

These problems could be viewed as refutations of the theory of evolution. Thus, if species evolved from other species, one should have been able to discover intermediate forms in the geological strata. But no such intermediate forms were discovered.[28] Therefore, the species did not evolve. (Was the solitary case of archeoptryx, which showed a link between birds and mammals, enough?) But the methodological convention dictates that the refutation must be sought in one of the auxiliary assumptions in the protective belt. And so it was. Darwin conjectured that the stratigraphic record must of necessity be poor, scattered, and inadequate.[29]

[28] This was a commonplace objection in the nineteenth century after 1859, and I cite just one example: St. George Jackson Mivart. The paleontological evidence, he said, in *On the Genesis of Species*, "is as yet against the modification of species by 'Natural Selection' alone, because not only are minutely transitional forms generally absent, but they are absent in cases where we might certainly *a priori* have expected them to be present," 128.

[29] Second example: If species evolved from one another, the amount of time required for evolution to take place from the unicellular organism in the pre-Cambrian to *Homo sapiens* should be incredibly large. However, said William Thomson (Lord Kelvin), the latest results in physics dictated that the history of the Earth was considerably shorter than the span of time required for evolution to occur. Therefore, the species did not evolve. Once again, refutation was sought in adjusting the claims of the protective belt rather than those of the negative heuristic.

Thomas Huxley argued that biology simply must work within the constraint of time provided by physics. So if the time interval was shorter than expected, evolution must have speeded up in yet-to-be-discovered ways. Thus, said Huxley: "Biology takes her time from geology. The only reason we have for believing in the slow rate of the change in living forms is the fact that they persist through a series of deposits which, geology informs us, have taken a long while to make. If the geological clock is wrong, all the naturalist will have to do is to modify his notions of the rapidity of change accordingly." "Geological Reform," 331.

A scientific research program is evaluated at appropriate intervals of time.[30] A program is in a theoretically progressive problem shift if it can make at least some independently testable predictions. A program is in an empirically progressive problem shift if it makes at least one independently testable prediction, and the prediction is verified. Otherwise, a program is in a degenerating problem shift. The convention of the methodology of research programs dictates that bold conjectures in an empirically progressive problem shift must be pursued, and those in a degenerating problem shift should be regarded as refuted and abandoned.

Now, imagine Lakatos maintaining meta-conjecture 2, namely, "The meta-conjecture 'The methodology of research program is a better guide in the field of science than Popper's methodology of trial and error' is a correct one, although this meta-conjecture, like Popper's, is also fallible." In this version of Lakatos's method, he does not insist on any inductive principle like the following: "If a theory is in an empirically progressive problem shift, then it must be true, or probably true, and will continue to be in an empirically progressive problem shift." Nor does he maintain that "If a theory is in a degenerating problem shift, then it must be false, or probably false, and will continue to lead to poor results if pursued." Thus, inductive links are denied. Nor does Lakatos attempt to *justify* any of these claims, since that would be akin to solving the problem of induction. All he need claim is that since his method is a contingent claim, it may not succeed. But then again his method, which is akin to Popper's six rules, may succeed after all.

What I have said of Popper and Lakatos is true of virtually every philosopher in the objectivist tradition. Thus, other philosophers

[30] No Popperian can afford to cavil with the notion of an appropriate interval of time as being too loose to be useful. For Popper has argued that unless we are dogmatic with respect to our theories, we shall never be able to determine their fruitfulness, but we must clearly not immunize our theories from falsification. So there is a period of time when we are urged to be dogmatic and a time when the theory should be given up. See Popper, *Objective Knowledge*, 30, and *Conjectures and Refutations*, 49. The hard core represents the part one ought to be dogmatic about, but there is no danger of immunization from falsification in the methodology of research programs. A reasonably clear condition, such as a research program in a degenerating problem shift, is stated under which a research program should be given up.

would maintain that if a society of scientists is to be rational, then it should be structured along the lines of *their* respective methods. This is hardly surprising given their own respective meta-conjectures. For instance, Putnam's inductive method of schemata might make a similar claim.[31] Putnam would maintain the following meta-conjecture 3: "The method of schemata is a better, if fallible, guide to discovering approximate truth than any extant method." This meta-conjecture too falls in with the objectivist view: A rational society must be structured along the lines of a single best available method, namely, Putnam's method.

It will now be a little easier to explain why a single-method solution is quite risky and too constraining. First, all philosophers in this tradition rightly maintain that any method is a contingent and fallible hypothesis of how scientific theories should be selected, which ones should be rejected, which ones should be modified, and so on. If the scientific society were founded on the single best available method – say, Popper's – then if the six rules are mistaken, objectively speaking, the scientific society would be led to a great impasse; it would be bedeviled by wrong guesses on theory selection, and the bad guesses would hamper the pace and growth of science and knowledge. Such a society, so structured, would be taking enormous risks.

Second, a society failing to make room for rival methods will not be able to compare the performance of competing methods and determine which methods should be rejected as inadequate, which ones can be improved, and which ones are good enough to be included in the structure of the society. Clearly, such a society would not be able to judge rival meta-conjectures, for the rivals are given no space to flourish. Thus, the structure of the society will be too constraining.

What if we have discovered the true method? One could argue that the true method will ensure that the scientific society will be an ideal one, and consequently that there is no necessity for multiplying methods. Given that our methods and meta-conjectures are fallible, the chances of our currently holding the true method are as good as the chances of our currently holding *the* true theory in any domain of science. What, then, if the selected method is utterly inadequate?

[31] See Sarkar, "Putnam's Schemata."

Or what if all our selected, competing methods are poor? Such a state of affairs would surely be disastrous for a scientific society. No solution to the problem of group rationality can avert this situation with a guarantee. But we can try to avoid the possibility. This brings us to the new problem of demarcation.

V. A New Problem of Demarcation

The old problem of demarcation was how to distinguish science from pseudoscience, not to be confused with the older problem of how to distinguish meaningful statements from nonmeaningful ones. Popper began with strong, pre-analytic intuitions about some theories. He believed that the Freudian theory of psychoanalysis and its ilk – astrology, palmistry, Marx's theory of history, and such – were pseudoscientific theories, pretenders to the status of science, and in fact characterizably different from Newton's theory of mechanics, Einstein's theory of general relativity, or the theories of the Ionians, Aristarchus, Kepler, and Galileo. As Popper emphasized in delineating the problem of demarcation, he was not interested in characterizing the properties of theories in virtue of which they were true, precise, mathematical, or measurable. He was rather interested in solving the following problem: "When should a theory be ranked as scientific?" or "Is there a criterion for the scientific character or status of a theory?"[32]

He solved the problem as follows. A theory is scientific if, and only if, it is empirically testable, while a theory is pseudoscientific if it lacks just such testability. A theory is testable if it forbids at least some state of affairs, whereas it is untestable if it fails to forbid any state of affairs. Popper then argued that all the theories he had initially put in the class of scientific theories were testable in his sense, and that all the theories in the class of pseudoscience were nontestable. He thereby claimed to have solved the problem of demarcation.[33]

Not everyone was persuaded. Some hinted that the problem of demarcation was not even a genuine philosophical problem, let alone

[32] Popper, *Conjectures and Refutations*, 33; emphasis dropped.

[33] See, especially, *Conjectures and Refutations*, Chapter 1; *Objective Knowledge*, Chapter 1; and *Realism and the Aim of Science*, Chapter 2.

the question whether Popper's solution to it was adequate. Consider, for example, what Hilary Putnam had to say. Popper had claimed that Darwin's theory of evolution is a successful metaphysical research program, but not a scientific program, on the grounds that it fails to satisfy the principle of falsification, the principle or method by which all sciences should be governed; thus, among disparate sciences, the principle of falsification exhibits methodological unity. Putnam, by contrast, argued for methodological disunity, claiming that the scientific status of Newton's theory of mechanics is underwritten by a method different from the one that confers scientific status upon Darwin's theory of evolution and natural selection.[34] "I have to ask," says Putnam,

> why on earth we should expect the sciences to have more than a family resemblance to one another? They all share a common heritage, in that they owe allegiance to a minimum of empiricism . . . ; they are conducted in a fallibilistic spirit; they frequently depend on very careful observation and/or experimentation . . . ; and they interact strongly with other disciplines recognized to be sciences. But there is no set of "essential" properties that all the sciences have in common.
>
> If evolution theory does not, taken as a whole, fit Popperian (or more broadly positivistic) accounts of science, there are other models it does fit.[35]

Two points are worthy of note, one relatively less important, with which I begin. First, Popper too owes allegiance to a minimum of empiricism (by implication, also to a dependence on careful observation and experimentation) and full-blooded fallibilism; furthermore, Popper too would commend strong interaction among the sciences (if only as a way of making theories and conjectures even more falsifiable).[36] But since observation and experimentation, fallibilism, and interaction among the sciences are under the direct control of falsification, what distinguishes these other methods, which allow evolutionary theory the status of science, from Popper's method or principle of falsification?

[34] Putnam, *Words and Life*, especially, 465–72.

[35] *Words and Life*, 471–72. For remarkably similar claims, with the same Wittgensteinian thrust, made much later (Putnam's were published in 1987), see John Dupré, *The Disorder of Things*, especially 229–33 and 342–43.

[36] ". . . all progress in science must proceed within a framework of scientific theories, some of which are criticized in the light of others." Popper, *Conjectures and Refutations*, 132.

Second, let us grant Putnam his central point: There are, and ought to be, different methods for treating distinct sciences. Are there then distinct methods for each scientific domain, a fresh model for each scientific discipline? If so, how do we eliminate competing methods in each domain? Or will there be several models fitting each discipline? What, then, is the notion of fit in this context? Might Velikovskians, Creationists, and Freudians defend their own disciplines by emerging with their own distinctive methods, each group claiming unique fit? Or how will we be able to tell when a discipline bears little or no family resemblances to disciplines regarded as sciences? Which resemblances are significant, anyway; and why?[37]

My purpose, however, in sketching this argument and counterargument on the old problem of demarcation is to use it as a preface to a new problem of demarcation that, I believe, is utterly important. Moreover, the old problem of demarcating science from pseudo-science cannot be regarded as solved until the new problem is (as I have already hinted). Let us call the new problem the problem of method demarcation. The problem can be stated thus: "How is a good method to be distinguished from one that is not?" or "How is a good method to be distinguished from a pseudo-method?" I too am not interested in characterizing the property of a method in virtue of which it is true, fruitful, precise, and so on. Rather, I wish to determine the property in virtue of which a method is labeled "good." Our task, by analogy, is: Just as testability is a property of a theory in virtue of which it is scientific, what is the property of a method in virtue of which it is good? Why this problem should be important is not far to seek: In the problem of group rationality, method is primary.

The problem is particularly significant to someone who thinks that the scientific character of a theory is dependent on having a good method. Where there is no good method, there is no way of determining what distinguishes science from pseudoscience. If that is true, then unquestionably the first task is to determine the property of a method in virtue of which it is good. And this is true of Popper. No delineation of Popper's solution to the problem of demarcation is adequate that

[37] In view of the foregoing, it is worth determining to what extent the controversy between Popper and Otto Neurath on the issue of method – delineated in a rather engaging way by Nancy Cartwright, Jordi Cat, Lola Fleck, and Thomas E. Uebel in *Otto Neurath*; see especially 202–8 – settles in favor of Neurath's two principles.

fails to consider method, or what Popper calls the second-order tradition. He repeatedly emphasized the importance of the second-order, critical tradition, which rendered much of, say, the Ionian output scientific and showed why later traditions, uncritical as they were, as in the Middle Ages, produced only unscientific theories. So criticisms of Popper's solution that show scientific theories to be untestable but do so by treating the theories in isolation from an ongoing critical tradition are far too simple and uninteresting.[38]

Yet Popper confused the two problems of demarcation. This is strikingly evident in the only passage I know in which he speaks of the old problem of demarcation but inadvertently formulates the new one:

> I often formulated my problem as one of distinguishing between a genuinely empirical method and a non-empirical or even a pseudo-empirical method – that is to say, a method which, although it appeals to observation and experiment, nevertheless does not come up to scientific standards. The latter method may be exemplified by astrology, with its stupendous mass of empirical evidence based on observation – on horoscopes and on biographies.[39]

Theories and methods are different. A theory is about the world and the world's causal structure and process; a method is about theories. On Popper's view, the difference is deeper still. A scientific theory is empirical, a method only conventional; so a theory can be true or verisimilar, but a method can be neither. Consequently, in the foregoing passage, we can understand what it is for astrology to be nonempirical: Although it amasses observations, it is not testable and therefore pseudoscientific or pseudo-empirical. However, one cannot speak of a method, or the second-order tradition, that underlies astrology in a similar fashion, for methods are not empirical, and hence not testable, so it is unnecessary to gather on their behalf a stupendous mass of empirical evidence based on observation. So, clearly, Popper's formulation of the problem of demarcation in this single passage is erroneous. It is so precisely because there are two problems, and not just one problem that can be characterized in two ways.

[38] Despite Adolf Grünbaum's many deep criticisms of Popper, some of that criticism is vitiated by his reluctance to consider the second-order tradition.

[39] Popper, *Conjectures and Refutations*, 33–4.

Consider the issue from the point of view of the objectivist solution to the problem of group rationality. A single best available method must govern the basic structure of a scientific society – but surely not *any* method. There must be two steps in the solution. First, there must be a solution to the problem of method demarcation, which will enable us to distinguish good methods from those that are not. Once that boundary line is drawn, then, second, we will be able to pick the best method from the class of good methods. It is *this* method that, on the objectivist view, should govern the basic form or structure of the scientific society.

The first step is crucial. For the objectivist view does not want any method to structure the scientific society. For instance, if at some stage in the history of science, some particular society of scientists – say, the society of phrenologists in the nineteenth century, the society of Franz Joseph Gall, George Combe, J. G. Spurzheim, and others – had no good methods at all, and the society picked the best available method from the class of methods that were not good to structure its society, then the objectivist view would not regard it as a rational society. The analogy with scientific theories is obvious. If we wish to pick the best scientific theory, we have two problems to solve. First, we divide all extant theories, in a given domain, into those that are scientific and those that are not. Then, second, we pick the best available theory from the class of scientific theories.

The problem of method demarcation is easier to illustrate and understand than to solve. Before attempting to solve the old problem of demarcation, we at least had a vague idea of what constituted a scientific theory and what an unscientific one. The following theories are regarded as scientific par excellence: Maxwell's theory of electromagnetism, Bohr's theory of the atom, the theory of plate tectonics, and Wallace's "On the Law which has Regulated the Introduction of New Species." Then there were the usual examples of theories that were pseudoscientific: palmistry and Paracelsian medicine. Hence, we had more than an inkling of what the solution to the problem of demarcation must accomplish.

Correspondingly, then, we should first give examples of good methods, and then examples of poor methods. Nothing more than intuition need serve as a starting point. Perhaps, so as not to beg the question, we should give examples of rational societies and then examples of irrational societies. Whether such societies are irrational because of their poor methods or for some other reason is left open. Our task is to come up with an adequate philosophical theory that will demarcate these societies into two mutually exclusive and exhaustive classes. Such a theory would in part enable us to explain why some societies are rational and others are not. *In the absence of a solution to the problem of method demarcation, we cannot have a completely adequate theory of group rationality, either from the objectivist point of view or any other point of view.* Let me draw, then, an example from the history of science to illustrate what, I think, was a society founded on a poor methodological tradition, followed by a cluster of instances of good methodological traditions.

VI. Illustration: The Herbalist Tradition

Consider the herbalist tradition. When William Turner published his *Libellus de re herbaria novus* (*The New Little Book about Plants*) in 1538, inaugurating English botany, the herbalist tradition was over seventeen hundred years old. Among the herbalists of the first century B.C. were Crateuas, Diocles, and Sextus Niger. But the chief herbalist, who lived in the following century, was Dioscorides. What was the aim of the herbalist? The tradition itself is clear enough, if not wholly clear in its aims; sometimes its aim was to classify plants as plants, but mostly it was to classify them from the point of view of their medicinal value. The aim is best summarized in the statement from the *Prologue of Der Gart der Gesundheit (The Garden of Good Health)*, a book by the fifteenth-century herbalist Peter Schoeffer: God, "Who has set us amidst... perils, has granted us a remedy with all manner of herbs, animals, and sundry other created things."[40] The issue before us is this: How did the group of herbalists – say, from Pliny the Elder to Turner, or, at any rate, most of them during that period – go about fulfilling their common aim of discovering cures and remedies for ailments and diseases?

The first significant fact about the herbalist tradition was that of copying from others. Oftentimes, therefore, a herbal was no more

[40] Frank J. Anderson, *An Illustrated History of the Herbals*, 96. I am much indebted to this marvelous book for the present illustration. I have also benefited from reading the relevant chapters and sections from Lynn Thorndike's massive *A History of Magic and Experimental Science*.

than a compendium of texts and illustrations gathered from a variety of contemporary or earlier herbals.[41]

The second feature of the herbalist tradition was that there was little or no theorizing. There was the mere collection of facts; the search, and the subsequent report of the search, were sufficiently random to be susceptible to no probing questions and to no common scheme of explanation. There was virtually no systematic theory involved, given that the listing of plants was more often than not entirely based on alphabetical order, and the text followed the train of rich illustrations in the book, and not vice versa. It is true that Dioscorides had complained in the famed *De Materia Medica* (no longer extant) against using the alphabetical method of classifying medicines because closely related plants would be separated, consequently making it difficult to discover the particular properties and operations of plants.[42]

The third, and the most significant, fact about the herbalist tradition was the lack of testability, or at least of probing into alleged facts, before stating them or copying them from other herbals. In short, there was no critical method. An example: Pliny reported a sea dragon that was a cross between a reptile, a butterfly, a fantailed goldfish, and a one-horned ape![43]

[41] Pliny copied from his predecessor; his successors copied from him; they were copied by those who came after them, and so on. For example, Dioscorides copied from Pliny; Macer Floridus copied from Dioscorides; 264 of the 474 chapters in *Le Grant Herbier* (authorship disputed) are borrowed from Matthaeus Platearius's *Circa Instans.*

[42] *De Materia Medica* overcame this problem to some extent in its treatment of aromatics, oils, ointments, and trees; milk and dairy products, cereals, and sharp herbs; roots, herbs, seeds, and juices; and finally, wines and metallic ores. However, the rationale of the Dioscoridean system was still missing. For instance, herbs that were used for diuretics, cathartics, and emetics, or for other medical purposes, were randomly distributed throughout the book. Likewise with Pliny the Elder: Of the thirty-seven books of his *Natural History*, sixteen were concerned with plants and eighteen with medicines and diseases. However, *Natural History* was a huge compendium of fact and fable, as was much of *De Materia Medica.*

[43] Here are some more examples. From *The Herbal of Apuleius*: The plant verbena was recommended for ulcers and swellings, and for dog, snake, and spider bites. When eaten, verbena was claimed to cure indigestion, relieve liver ailments, and remove bladder stones. Macer, in *De viribus herbarum*, suggested "bruising" the flowers and leaves of the plant groundsel in wine and applying the mixture to tumors of the rectum and testicles. The nun Hildegarde of Bingen said in her work *Physica* that in order to enable women to overcome difficulty of labor, a lion's heart should be placed, but only for a short interval of time, on the navel of the pregnant woman.

Far from being isolated examples, it is fair to say that such claims characterized the herbalists' practice. But there was a happy exception: Hieronymus Bock, the German author of *Kreuterbuch* (1539). Dioscorides had claimed that ferns produced no seeds. The German peasants thought otherwise, but they nonetheless surrounded the sowing and harvesting of ferns with magical rites. Bock settled the matter by investigating the plant, without resorting to these rites, on four successive Midsummer Eves, when ferns formed seeds; he concluded that Dioscorides was wrong. With Bock, also, the idea of system came into existence together with the ideas of plant communities, plant description, and plant environments.

Why was the society of the herbalists so poor, so irrational? One answer is that their method was poor. It is very difficult to discern precisely what the method was, or what the methods were, that were used over hundreds of years. However, I shall aver that, intuitively, whatever these methods were, they were not good, by showing that what the scientists chose to do with their theories was entirely implausible, irrational. In other words, I shall focus on the tradition of this group of herbalists and infer certain things about its tacit methodological tradition. Our problem of method demarcation, then, is to distinguish the earlier practices of the herbalists from the later ones, like Bock's in the mid sixteenth century.

Notice: I am not merely suggesting that the theories of the herbalists were not put to an experimental test, as indeed they could have been, but rather that there was *no* critical discussion at all.[44] Prior to Schoeffer's *Der Gart*, no illustrations of herbals were made by observing nature; blind copying of the texts and illustrations was the *rule*. So the possibility of correcting illustrations was largely forsaken. The notion of critical discussion or criticism is broader than the notion of an experimental or empirical test. There were no questions like: Could this be right? Could placing a lion's heart on a woman's navel while she was in labor ease her pains? If so, how did it do so? Is this theory inadequate? If so, why? Which theory seems promising to pursue? Such questions were rarely asked, primarily, I suspect, because there were no clear

[44] This paragraph is in harmony with the objectivist's view of critical method. See also Sarkar, *A Theory of Method*, Chapter 2, section III, particularly the chart on p. 43 and the discussion connected with it.

methods in this field to which one could appeal, or the methods were too uncritical or too vague. The absence of a method, or it being ill-defined, was a reason why science suffered and why medicine, botany, pharmacology, and natural history were not yet distinct disciplines, and why the herbalist tradition tried to embrace them all. No historian, to my knowledge, has culled out the method involved or invoked in the society of the herbalists. From their practices, we infer that it was a poor method.

By contrast, consider the society of physicists in the time of Michael Faraday and Max Plank, of biologists in the time of Asa Gray and Richard Owen, of botanists in the time of John Gray and Carl Linnaeus, of geologists in the time of Charles Lyell and James Hutton, of chemists in the time of John Dalton and Lavoisier, and so on. We have studies of these societies from virtually every stance. Philosophers of science have looked at them from the point of view of the theories they produced. They have asked if these theories showed signs of progress, if their terms had common meaning and reference, if earlier theories were limiting cases of later theories. Social historians have asked how the political structure or the economic struggle affected the content of those sciences. But rarely have we looked at the structure of the society as defined by its methods. If and when such a task is performed, then we shall have enough information to judge the adequacy of the solution to the problem of method demarcation. An adequate solution will show us why the latter types of societies are different and distinct from the types of societies that characterized the herbalists. *Then* shall we understand better the make-up of rational societies of scientists.

VII. The Scare of Saint-Simon

"I wanted to try, like everybody else, to systematize the philosophy of God. I wanted to descend in succession from the phenomenon universe to the phenomenon solar system, from there to the terrestrial phenomena, and finally to the study of species, considered as a dependency of the sublunar phenomena, and to deduce from this study the laws of social organization, the original and essential object of my research."[45] To go along with this immodest aim were immodest assets:

[45] See Keith Taylor, *Henri Saint-Simon (1760–1825)*, 65.

"I believe, gentlemen, that I have found an encyclopedic conception better than that of Bacon, a conception of the universe better than that of Newton and a method better than that of Locke."[46] So said Claude Henri de Rouvroy, comte de Saint-Simon.

To put these ideas into practice he needed no ordinary political power. The emperor, Napoleon, had to be sought, his kingdom extended. Thus, "The Emperor is the scientific as well as the political head of the community." But the emperor cannot himself bring the envisioned society of scientists into being. He will need "a scientific lieutenant capable of understanding his projects: he needs another Descartes." With Saint-Simon in the role of Descartes, the emperor will engineer into existence a "scientific monument of a dimension and magnificence which cannot be equalled by any of his successors." And so Saint-Simon invited some sixty of the greatest scientists of the age, by sending them a copy of his *Memoire sur la Science de l'Homme*, to come to Paris and labor as a group under Napoleon's regime to produce a compendium of all knowledge.[47]

Here are embodied the central tenets of nineteenth-century positivism, which was to influence Augustin Thierry and Auguste Comte, Vladimir Lenin and, to some extent, Karl Marx. Saint-Simon set out to do for the social sciences what Newton had done for physics. The history of man is determined by laws that are empirical and discoverable; these laws are writ large. It is the task of the social scientist to discover the laws governing the social phenomena by carefully studying social, political, and scientific history. History has a direction and a purpose. The historical process can be clearly and consciously directed toward consciously adopted ends, ends that are in harmony with the historical process. When history is allowed to take its natural course, what results is a waste of talent and resources.

Once the laws of the social sciences are discovered, then society can be organized so as to be in step with sociological laws and to reap the full benefits of proper direction, constraints, and method. It is the task of the philosopher to use these laws, to harness them for worthy social and political causes as well as for the cause of science. He must minimize the loss and waste by organizing the society of scientists

[46] See Mathurin Dondo, *The French Faust: Henri de Saint-Simon*, 109.
[47] See Dondo, *The French Faust: Henri de Saint-Simon*, especially Chapter 7.

in light of his knowledge of the past and the relevant social laws. Consequently, science will proceed unhampered and apace. When full details of how the society of scientists is to be structured are specified, there will be little room for chance or freedom to hamper the growth of science. The value of the social process transcends the value of single individuals. Thus the freedom of individuals can and must be restricted for the sake of higher social ends, not exactly the skeptic's view (nor mine). Saint-Simon accepted the idea of lack of liberty: "I cannot conceive of association without government of someone."[48] It would follow, therefore, that the association of scientists would need a governor: Saint-Simon.

If this frightening prospect – the scare of Saint-Simon – is what a rational society should be, and how it should be engineered into being, then let us have less of rationality.[49] Perhaps the problem of group rationality should be feared. For a solution to that problem, a skeptic might say, can only generate a Saint-Simon-type response in the field of science. A solution to the problem will be regarded as a blueprint for a rational society. Social engineers will be summoned, individual liberties will be curtailed, and massive changes and transformations will be undertaken until the envisaged society of scientists is a reality. But, the skeptic might continue, there is no surety of outcome: Sweeping changes will not, and cannot, leave notes behind about what the future will hold. If changes are executed on a sweeping scale, we will be unable to take precautionary measures. We will be swept aside or swept away by our own undertakings. Things will be out of control, they will take on a momentum of their own, and events will rush on to unforseen or unforeseeable ends – most likely, undesirable ends. Traditional ways of doing science will be done away with, and new methods will be introduced, with little knowledge of the consequences of doing either of these things. The society of scientists will face instability and will be in danger of losing whatever good science it possesses, whose growth and development was the work of ages. So a skeptic might say.

[48] To which his student Augustin Thierry replied that he "could not conceive of association without liberty." See F. H. A. Hayek, *The Counter-Revolution of Science*, 234.

[49] On a smaller scale, in England, Thomas Huxley and Joseph Hooker violently opposed Richard Owen's ambition to institutionalize centralization by having a museum built to Owen's specifications and with Owen at its head, thus putting the fate of all natural history sciences in the hands of one man; see Browne, *Charles Darwin: The Power of Place*, 98.

But must the solution, any solution, to the problem of group rationality mimic the style of Saint-Simon? Consider: An engineer's problem is nearly always well defined. His task may be to build a bridge, a canal, or an airplane to reasonably precise specifications. He knows the limits of the available resources, time, and money, the geographical boundary, and the tools and instruments at his disposal. His solution is a solution to the problem only when it satisfies those limits and constraints. Anything too large, too small, too time-consuming, or too expensive is not a solution, and he must return to the drawing board. Given the considerable success of the engineer, it is only natural that we use the engineer as the model and guide for executing other tasks and for solving other problems.

Thus, there should be a social engineer. The tasks and goals of the social engineer are similar to those of the engineer, except they are bigger, less manageable, and incredibly more complex. The task might be this: We come with a blueprint for a just and a good society, a social plan, and the social engineer is to put the blueprint into practice by staying within the limits of the national resources. He is free to change and alter significant parts and practices of the society in order to effect large-scale changes, but the aim of bringing about large-scale changes is to push or pull the society away from its current drift into a mold specified by the utopian blueprint. This entails that the social engineer must be able to predict the multiplicity of effects that will occur when myriad parts and processes, institutions and traditions, customs and practices of the society are simultaneously changed or altered. He must be able to foretell whether these changes will collectively lead in the direction one hopes to go. His performance will be judged by how successfully he is able to transform the society into a just and a good society, how well he can bring to life the social plan as a whole, not merely some of its parts.

In the liberal, objectivist view, the key phrase is not *social planning*,[50] it is rather *piecemeal social engineering*. Saint-Simon-type changes are

[50] Social planning on a large scale was unsparingly opposed by three distinguished men of our times: Friedrich Hayek, Karl Popper, and Edward Shils, all of whom early in the twentieth century taught at the London School of Economics and Political Science. In particular, see Hayek's *The Counter-Revolution of Science*; Popper's *The Open Society and Its Enemies*; and Shils's *Tradition*. When I read Shils's *Tradition*, I laugh at the use of the notion of tradition by philosophers of science, this one included. So rich is the concept and so vast its utility that only Shils's learning could portray it adequately.

vast, widespread, and sweeping; by contrast, the type of changes recommended by the objectivist are small, circumspect, and welldefined; such changes are part of the famed notion of *anticipating unintended consequences*. The virtue of this view is that piecemeal steps enable us to check, control, or correct the plans and policies by studying the slow, streaming effects that flow out of putting the plan into practice, and by modifying these effects or rendering them harmless. The goals to be achieved are equal liberty, lessening of pain and misery, and elimination or reduction of unemployment. The goal is not to produce goods and happiness; it is rather to remove the obstacles and burdens of poverty, lack of opportunity, and ignorance.[51] The goal is not that of positive utilitarianism, but rather that of negative utilitarianism.

Now, distinguish the plan from the process. Once that distinction is drawn, the objectivist view need not fear the skeptic's argument. Suppose the right solution to the problem of group rationality is the objectivist view: A scientific society is rational if, and only if, it is structured along the lines of the best available method, given the goal of truth or verisimilitude. The method is unique in that it pushes science in one direction, in terms of both depth and surface detail, with the likelihood that the goal of truth or verisimilitude may be fulfilled. This is the plan, the blueprint. Does it follow that the process of bringing about a rational society of scientists must be sweeping and total? Could not the process be one advocated by piecemeal social engineering? If a scientific society is to be founded on a single method, then slow, but deliberate, changes must be initiated in that society that over a period of time will eliminate several methods hitherto structuring the society. There is nothing in the solution that suggests that elimination of old methods must be immediate and simultaneous or, in other words, that the restructuring of the scientific society must be on a massive scale, in one fell swoop. There are simpler ways of achieving a revolution.

[51] For what is now a classic statement on how and why economic development should be seen primarily in terms of furthering the cause of freedom, expanding freedoms we have reason to value, freedom as opportunity as well as process, and not merely in terms of utility, procedural liberty, or real income, see Amartya Sen, *Development as Freedom*. It is worth noting that in the liberal, objectivist view of Carl Menger, Friederich Hayek, and Popper, the idea that the social sciences should be concerned with anticipating unintended consequences lies at the core of that objectivist view. Sen has been a relentless critic of the notion of anticipating unintended consequences; see, for example, *Development as Freedom*, 254–7.

There are several questions that will need to be answered. What are the current theories of science, the auxiliary hypotheses, experiments, and interesting theoretical problems? Which current methods speak to these? What, if any, redundancies exist in the methods? What will be the unintended consequences for science in initiating this change? How will the science of the new group be analyzed and evaluated? Which among the current methods seem to be most promising? Keeping in mind the Williams problem,[52] what social, political, and economic conditions are necessary to eliminate various methods? How will the fruitfulness of the transformed group for the rest of society be measured? It is only after these questions are answered, at least tentatively, that a new method may be introduced into the society of scientists. This is the folk wisdom of the liberal view. Consequently, in terms of process, the procedure is not different from the one advocated by the liberal view. At any rate, the process is clearly not designed to delight Saint-Simon.

Suppose, now, that a skeptic were to counter with this: "No one in the history of science has even once been able to experimentally introduce into an ongoing scientific community a single method to see if the method will produce useful results. The methods that are used in the actual practice of science are in large part a matter of circumstance and accident; for the rest, they are determined by what the immediate past happened to be. When a new method is introduced, it hardly takes the form of a conscious, explicit, and deliberate attempt to introduce it into the scientific community. Karl Marx says, 'Men make their own history, but they do not make it just as they please; they do not make it under circumstances chosen by themselves, but under circumstances directly encountered, given and transmitted from the past.'[53] So in science: Men make their own science, but they do not make it just as they please. There is first the immediate past of the scientists, and

[52] For the objectivist view, it would read thus: What is the ideal social order (Social Utopia) presupposed by the objectivist view of group rationality? Is that Social Utopia pliable and morally justified? What is the ideal scientific society structuring the practice of scientists (Scientific Utopia) presupposed by the objectivist view? Is that Scientific Utopia realistic, and how successful would the scientific practice be in that utopia? Can the Social Utopia lie in cohesion with the Scientific Utopia? See Chapter 1, section III, this volume.

[53] Karl Marx, *The Eighteenth Brumaire of Louis Bonaparte*, 15.

then there is the basic structure of their society. That structure is not resilient; while it influences the scientists in producing the theories they produce, its transformation, surely, is beyond the will or control of solitary scientists.

"As we consider more advanced and contemporary science, the issues, practices, and problems of that science and the manner in which high science is financed are so enormously complex and convoluted that methods (save in the bumbling social sciences) plainly play no role at all. The chance of finding scientists who are willing to form a subgroup on the basis of a new method in order to determine experimentally the efficacy of the new method is simply too small, if not zero. The risk to careers would be too great, not to mention the loss of prestige, power, and time. So this prattle about group rationality and the possibility of engineering a rational society into existence by piecemeal procedures is just that – prattle."

How shall the skeptic be answered? There are three problems in all. The first is *the problem of understanding*. Of a particular scientific group we may wish to ask what appears to be a straightforward empirical question: How does this scientific society function? What is its structure? How does the political, social, and economic structure in which the society is embedded affect the origin, pace, and growth of science? We are not interested in evaluating the society, we merely wish to understand its form and function. The second problem is *the normative problem*: How ought a scientific society to be structured? The answer to that question would bring forth a blueprint for a Scientific Utopia and, given the Williams problem, would raise the question about the Social Utopia that will be needed to engineer the Scientific Utopia into existence. Of a particular scientific group we may wish to ask whether it is structured as it should be. In short, we wish to evaluate it. Finally, there is *the problem of strategy*. Given the answer to the normative problem, we may wish to ask: How shall we go about designing a scientific society that meets the specifications outlined in the answer (keeping in mind justice and other such things)? The skeptic was portrayed as being especially skeptical about answering the third question.

The normative problem is cardinal. Its link to the problem of strategy is easily established. If there is no solution to the normative problem, one cannot go about creating a rational society, for one would not know *what* one ought to create, whether one ought to create a society of scientists based on no method, a single method, or a few methods. This is not to suggest that the problem of strategy is solved once one has a solution to the normative problem – far from it. To build a bridge in a particular geographical location, one needs to take into account a host of contingent factors, such as landscape and seascape, available manpower and resources, none of which is given in the laws of physics, electricity and magnetism, or quantum mechanics. In a similar manner, the problem of strategy must take into account a host of contingent historical factors, such as the particular scientific problems the scientists wish to solve, their experimental needs, available resources, scientific personnel, and so on, in order to structure a society to the specified form; and once again, none of these contingent factors is given in the theory of group rationality.

The normative problem is also closely linked to the problem of understanding. A theory of group rationality will provide the necessary framework in which to answer empirical questions, such as: What were the effects of the anti-vivisectionist movement on physiology in nineteenth-century England? What hindered the growth of medicine in the Paracelsian period? What led to the formation of geological subgroups in Germany in the nineteenth century? These questions presuppose answers to questions like: What is a subgroup? What role and function does it have in a scientific group? What is science? What counts as progress, backwardness, or hindrance of a subgroup? of the group as a whole? These are precisely the questions whose answers lie in a normative theory of group rationality. Indeed, it is more than likely that with the development of theories of group rationality we shall be led to ask empirical questions we had never hitherto considered, and we shall then see the history of science under new and surprising aspects. Hence, my answer to the skeptic is that a theory of group rationality will enhance our understanding of science and the history of science. If it did nothing else, that would be reason enough.

Consider some of the traditional problems in philosophy of science. What is a scientific law? What is explanation? When is one theory reducible to another? What is the relation between theory and experiment, between science and metaphysics? Under what conditions should a theory be accepted? These are problems whose answers enhance our understanding; yet their role in the problem of strategy is doubtful. I should argue that the problem of group rationality is an additional problem on the standard list of problems, and that a theory that solves that problem will deepen our understanding of science and the history of science. Then the proper answer to the skeptic should be, "We do not wish to change scientific practice, simply to understand it."

This should be answer enough, but I do not wish to rest there. Consider social and political philosophy. Here we deal with questions like: What are individual rights? What are just institutions? What are fair obligations and duties? What, if any, are the permissible forms of taxation? A theory of justice will answer these questions, and our hope is to introduce the claims of a correct theory of justice into the ongoing practice of the society. The aim is to change the society, if we can. Often, we are unable to do so. No political philosopher says, "Here is a theory of justice; it is an ideal theory that could not be put into practice. It is only meant to deepen your understanding of the issues of social justice." One might say, rather, "Here is a theory of justice; it is an ideal theory. I do not quite know how to translate these precepts into practice. For a successful translation depends on having knowledge of a vast number of factors, the desire to take time off to see the changes through, and an appropriate political base from which to act. These I do not have. So I leave the introduction of a just society in accordance with my conception of justice to those who are better placed than I at implementing the conception or plan."[54] But if it is not improper for a political philosopher to say this – and it is not – why is it objectionable in a philosopher of science who proposes a plan for constructing a rational society of scientists (a Scientific Utopia)?

[54] "Look, if I thought that if I really went and spent the next few years devoting all of my energies to propagating libertarianism, that then we would have a libertarian society, I would certainly go out and do it. I think it is very important to have a libertarian society. But I am doing some expected-value calculation and weighting of the importance by the difference in probability that I think my activities would make, and I guess I do not think it would make all that much difference. . . . It's just that I don't think I have anything especially interesting or illuminating to say about how to get there from here. I would hope other people would be better at this than I.

"And now if there was a libertarian society, I would not choose to be a political figure in a libertarian society. So why should I be a prisoner of the time I am born in?" Robert Nozick, "Persona Grata: An Interview with Robert Nozick," 3 and 6.

8

Putnam, Individual Rationality, and Peirce's Puzzle

We have analyzed the problem of group rationality and outlined four theories – the skeptical view, the two versions of the subjectivist view, and the objectivist view – that address that problem. These views focused on the group of scientists, not on the individual scientist; in particular, they focused on group rationality rather than on individual rationality. It is imperative that we examine the notion of the rationality of an individual scientist to see what bearing, if any, it would have on group rationality. A closer examination of individual rationality might unhappily reveal that it – individual rationality – rests on rather shaky grounds. Perhaps not. Given the fundamental work of Hilary Putnam on this notion, we can no longer be sanguine about it. If the notion of individual rationality is suspect, or utterly unclear, our task would become enormously complicated. For how could a shaky, unsettled notion of individual rationality provide what scientists need – let alone provide all the scientists the same reason – to cohere into a group?[1]

Drawing upon the later philosophy of Putnam, I show in the first two sections that analogous to the problem of what moral vision should be harnessed to civil society (one to which Putnam offers a solution) is the problem of what scientific image ought to be wedded to the society of scientists. These sections will show that democracy is, at best, a necessary condition in which a rational society of scientists can flourish, and they will draw a much-needed distinction between Social Utopia

[1] See, especially, Chapter 6, section IV, this volume.

and Scientific Utopia. In section III, I shall consider Putnam's view on the fact/value distinction, historical knowledge, scientific method, conceptual relativity, and different kinds of reasons – scientific, ethical, aesthetic, and juridical – and show the consequences of these claims for the problem of group rationality. Section IV centers upon a powerful puzzle, Peirce's puzzle, that Putnam has reintroduced into the literature; I aim to show its keen relevance to both individual and group rationality. Section V attempts to show that Putnam's position ultimately leads to relativism (his protestations notwithstanding);[2] and the final section shows how he might, perhaps, forestall it.

These sections, please note, are fairly closely crisscrossed, intertwined. It would be a stretch to say I am merely sketching a theory. It would be closer to the truth to say that I am listing a set of problems – perhaps uniquely viewed – that a theorist of group rationality would need to solve if he is to have any hope of offering even a thin theory of group rationality.

I. Democracy and Group Rationality

L. Conradt's and T. J. Roper's "Group Decision-making in Animals," which appeared in *Nature* in 2003, argued that groups of animals often have to make collective decisions about which activities to perform, when to perform them, and the direction of travel. Given how little we know about how group decisions are made among nonhuman animals, they offered a testable model. Assuming two decision-making processes – namely, despotic and democratic – their model showed that under most conditions, the cost for the group as a whole as well for the subordinate members of the group is greater under the despotic process than under a democratic one. Only when the group size is small and information difference large does it pay for the group to be run by a despot. In all other cases, democratic decision making has bigger payoff because it tends to result in less extreme decisions, "rather than because each individual has an influence on the decision *per se*."[3] Assuming fitness is what is being sought, the authors conclude that in

[2] I take it as a given that Putnam has no taste for relativism; for example, see Hilary Putnam, "Why Reason Can't Be Naturalized," in his *Realism and Reason*.

[3] L. Conradt and T. J. Roper, "Group Decision-making in Animals," 155.

the nonhuman animal kingdom, democracy would be observed to be widespread.

Suppose a counterfactual: In the human animal kingdom, democracy is widespread. What significance would this have for scientists wanting to rationally structure their societies? Answering in terms of fitness would be insufficient at best, irrelevant at worst – as scientists are seeking to structure their societies rationally because they wish to seek truth or verisimilitude, not in order to increase their biological fitness.[4]

Consider this from Putnam:

> For Peirce and Dewey, inquiry is cooperative human interaction with an environment; and both aspects, the active intervention, the active manipulation of the environment, and the cooperation with other human beings, are vital. . . . For the pragmatists, the model is a *group* of inquirers tying to produce good ideas and trying to test them to see which ones have value.[5]

For the pragmatist, a method is not some algorithmic procedure; rather, it consists of maxims. So that:

> . . . when one human being in isolation tries to interpret even the best maxims for himself and does not allow others to criticize the way in which he or she interprets those maxims, or the way in which he or she applies them, then the kind of "certainty" that results is *in practice* fatally tainted with subjectivity. The introduction of new ideas for testing likewise depends on cooperation. . . . Cooperation is necessary both for the formation of ideas and for their rational testing.[6]

> What I have just offered is, in part, an *instrumental* justification of the democratization of inquiry.[7] . . . The claim, then, is this: Democracy is not just one form of social life among other workable forms of social life; it is the precondition for the full application of intelligence to the solution of social problems.[8]

[4] The oft-observed point: Scientists may have a well-confirmed theory predicting their own demise.

[5] Hilary Putnam, *Pragmatism: An Open Question*, 70, 71.

[6] Ibid., 71–2.

[7] Ibid., 73.

[8] Hilary Putnam, *Renewing Philosophy*, 180. For a useful analysis of the importance of Putnam's idea of a democratic ethical community, and some interesting criticisms of Putnam's views about objectivity in ethical and political disputes, see Richard J. Bernstein, "The Pragmatic Turn: The Entanglement of Fact and Value," especially 261–4.

Let us concede that democracy is at least a necessary condition of social life[9] in order for a scientific society to flourish;[10] this would still not be enough to tell scientists *in* the condition of democracy *how* to structure their societies rationally (assuming it to be a kind of social problem). This begs for a distinction between Social Utopia (democracy, say) – a utopia in which the scientists in the group have no conflicts about what things one ought to ethically, aesthetically, or religiously value – and Scientific Utopia, a state in which scientists in the group act in accordance with what a normative theory of group rationality dictates. Clearly, a Social Utopia will still not solve the problem of how the group of scientists should structure itself; in short, such scientists may differ on their Scientific Utopia. For this simple reason: Each of the views we have examined thus far – namely, the skeptical view, the two versions of the subjectivist view, and the objectivist view – can flourish in a democratic society. If it is false to say – and it seems patently false – that each such structure would flourish (in terms of truth and verisimilitude) just as well any other, we would have on our hands the task of evaluating the comparative worth of these four rivals

[9] So much so that "the society which is not democratic is in a certain way ill." Putnam, *Renewing Philosophy*, 182.

[10] It would be nice to have even a sketch of an argument for establishing democracy as a necessary condition for solving the problem of group rationality. Following John Dewey, Putnam offers this: Philosophy has no privileged access to its own special brand of truth or goodness, something that can be known a priori. The business of democracy is a messy business, no less than the business of defending its value. Unless we engage in politics, there is no way of knowing what constitutes our natures, talents, and capacities. (This is remarkably Kantian, I think.) Then, when we know our natures and what we need in order to flourish, we can defend the value of democracy.

As we engage in social and political life, we discover what stunts our growth, restricts our capacities. For Dewey, one noteworthy factor is *inequality*; he was certain that it narrows our possibilities. Massive inequalities engender egregious limitations, which restrict the influence the dispossessed have in their own society; this, in turn, has serious detrimental effects on that very society. Learning this is a matter of experience, not a result of armchair reflection. Nor is the insight a result of seeing things from a culminating "right" point of view. There are no final answers, for Dewey, inasmuch as inquiry is an ongoing, limitless process. See Putnam, *Renewing Philosophy*, especially 186–90.

Adding these conditions to the problem of group rationality, we get something like this: No group of scientists can be successful if there are serious inequalities of influence; there will be too much talent going to waste. There will be final answers neither about where the group ought ultimately to head nor about how it ought to organize. Answers to these questions will evolve as scientists engage in mutual consultation on the issues they find pressing.

in a democratic society. This worth is to be judged in terms of how fecund each rival would prove to be in generating true or verisimilar theories, and we would then have to compare their relative fecundity. It is clear that democracy does not offer a significant answer to the question, "How should scientists rationally structure their societies?" That task must squarely fall on the shoulders of philosophers.

II. Moral Images, Scientific Images

In Lecture III: "Equality and Our Image of the Modern World," in *The Many Faces of Realism*, Putnam argues in favor of equality and freedom. While Kantian in spirit, but not entirely Kantian,[11] drawing upon the theories of Jürgen Habermas and Karl-Otto Apel, and taking a stand against the moral theories of the medievalists (Aquinas among them, for sure, and Alisdair MacIntyre, with their talk about essences and functions), Putnam argues that the task of providing a moral image of the world[12] is at least as important a task as determining rights, duties, and virtues – a preoccupation of most moral philosophers. To understand what constitutes a moral image of the world, consider Putnam's account of the evolution of rights.

The value of Equality, says Putnam, was first introduced into the culture of the West as a religious notion; subsequently, it lost its religious

[11] According to Putnam, Kant's misconception was not about truth; rather, it was his belief that moral philosophy is impossible without a transcendental guarantee, a guarantee furnished only by the noumena – hence the postulation by Kant of the noumena alongside the phenomena; see Putnam, *The Many Faces of Realism*, 42. In what follows, I make no effort at being complete and exhaustive; I am concerned neither about the adequacy of Putnam's moral philosophy nor about his representation of Kant. The footnotes echo some of my reservations. I offer only an outline of Putnam's views that serves my purpose.

My purpose is to extract from Putnam what is useful for what concerns us here, namely, the problem of group rationality. For a splendidly clear account of the moral point of *The Many Faces of Realism*, see Ruth Anna Putnam, "Doing What One Ought to Do." In that paper, Ruth Anna Putnam explains what it is to have a moral image of the world and how that image is to be made effective; how to bridge the gap between knowledge and moral action; and why a sharp distinction between a hypothetical and a categorical imperative is not sustainable. She even reflects briefly on scientists, a society of scientists, and scientists' image of science, matters that concern us here; see Putnam, "Doing What One Ought To Do," 281–2.

[12] Putnam attributes this crucial idea to Dieter Henrich's lectures on Kant at Harvard; see Putnam, *The Many Faces of Realism*, 89, footnote 8.

roots and became something mysterious, and people paid it deference out of sentimentality. Or they grievously attacked the value of equality, as did Nietzsche. Then secularism succeeded in the human arena, as religion stood aside, and equality came to be understood as the rights each human has, these rights being the same for all. Putnam encapsulates these claims in his three fundamental principles, thus:

I. There is something about human beings, some aspect of which is of incomparable significance, with respect to which all human beings are equal, no matter how unequal they may be in talents, achievements, social contribution, etc.[13]

II. Even those who are least talented, or whose achievements are the least, or whose contribution to the society is least, are deserving of respect.

III. Everyone's happiness or suffering is of equal prima facie moral importance.[14]

But with Kant, according to Putnam, equality underwent a remarkable metamorphosis. "One thing, however, is clear, in the traditional formulations, theistic and secular alike," says Putnam, "the value of equality does not have much to do with individual *freedom.* What I am going to suggest is that Kant offered a radically new way of giving content to the notion of equality, a way that builds liberty into equality."[15] To understand this radical Kantian way, one has to draw the familiar distinction between autonomy and heteronomy. In heteronomy, one is being legislated to by another; in autonomy, one self-legislates. What, then, is it to self-legislate in the moral sphere? Kant joins together free will, autonomy, and the universal law, and claims that just as physical laws govern the physical world, so also the moral law governs the free will. To act freely is to act in accordance with the moral law, and to act in accordance with the moral law is to act autonomously; any other act would be heteronomous.[16]

[13] This would have to be tidied up, but since my direct concern here is not with moral, social, or political philosophy, I simply refer the reader interested in the caveats to my paper "Kant: Let Us Compare."

[14] Putnam, *The Many Faces of Realism*, 45. These three passages from Putnam were originally in bold type.

[15] Ibid., 46.

[16] "... so freedom, although it is not a property of the will in accordance with natural laws, is not for that reason lawless but must instead be a causality in accordance with immutable laws but of a special kind; for otherwise a free will would be an absurdity." Kant, *Groundwork of the Metaphysics of Morals*, 4:446.

One might argue that this does not differ much from the medievalist's view of morality. In that view, once one knows what one's function, essence, or *ergon* is, one then employs one's will and reason to act in accordance with that knowledge. This can also be expressed as saying that once we know what happiness is – happiness in its full Aristotelian dress – we then employ our will and reason to succeed in our lives at that Aristotelian task. This will not work, says Putnam: Kant has reminded us that there are very many different notions of happiness, not just the Aristotelian. Hence, we will not be able to build a universal ethics on a notion susceptible to so many interpretations. This is the sort of Kantian argument that harvests Putnam's approval.[17]

Let us suppose this Kantian argument is unsound. Suppose that we could reason our way to *knowing* our *ergon* or happiness. This would be quite undesirable, thinks Putnam. Kant had famously argued in *Religion within the Boundaries of Mere Reason* that it is a *good* thing that we do not have a proof for a fundamental moral proposition. If, *per impossible,* we did, it would give rise to intolerance, atrocities.[18] However, if this is true, it has a profound implication, not covered by traditional theories. The implication is that, in the absence of proof, we must allow for a multiplicity of moral visions in our (or at least in secular) society.[19]

[17] Putnam's endorsement of this argument notwithstanding, it is worth asking how it coheres with the multiplicity of moral images Putnam also advocates (more on this later).

[18] The medievalist might resist such a conclusion. If the principle of toleration is a principle their reason has discovered as it has the fundamental moral proposition in question (whatever it is), why might the medievalist not show benign indifference to those who have not yet reasoned themselves to that cardinal moral proposition? I see no a priori necessity in the claim that the discovery of a proof for the moral proposition would lead to intolerance, bigotry. It might even be possible that the moral proposition (whatever it is) itself entails the principle of toleration.

[19] "What we require in moral philosophy is, first and foremost, a moral image of the world, or rather – since I am more of a pluralist than Kant – a number of complementary moral images of the world." Putnam, *The Many Faces of Realism,* 52. Might one lay a charge of ethical relativism against Putnam? *Either*: The content of the moral image of the world is *so* open-ended that it is hard to tell what might be excluded, if anything. "Kant's own answer is that the only content on which we are justified in relying is our right to *hope,*" says Putnam; see ibid., 50. But, then, surely, even a Nietzschean might hope. *Or*: The moral image of the world is not open-ended, but has well-defined boundaries. Admittedly, the boundaries cannot be those drawn by Kant, since those boundaries would exclude much Putnam would have us preserve or include. Perhaps, then, it is an open question in Putnam's moral philosophy just how and where to place the perimeter within which moral images must reside.

And this, with the moral image supplying the backdrop, is the sum force of propositions (I), (II), and (III).

"My claim, then," says Putnam,

> is that Kant is doing what he would have called 'philosophical anthropology', or providing what one might call a *moral image of the world*. . . . he is also, and most importantly, providing a moral image of the world which *inspires* these, and without which they don't make sense. A moral image, in the sense in which I am using the term, is not a declaration that this or that is a virtue, or that this or that is what one ought to do; it is rather a picture of how our virtues and ideals hang together with one another and of what they have to do with the position we are in.[20]

For what I am about, *this* is the central insight of Putnam I want to advertise.

Here, then, is the parallel I wish to draw: One of our central tasks as theorists of group rationality is to provide what one might call a *scientific image of the world* – to provide a scientific image of the world that *inspires* scientists to agree to form a group under that image, and without which neither they nor their activities would make much sense. The scientific image, in this sense, is not about this or that science, this or that technology, or this or that methodology. It is rather a *picture* of how that society of scientists and their theories, methods, and ideals hang together, and of what they ought to do, given the group they are in.

This is what Putnam's work has inspired. It is, therefore, worth asking, "What corresponds, or is analogous, to (I), (II), and (III), in the case of a society of scientists, which would display the scientific image of the world?" No final answers will be provided here – in fact, we shall see how difficult the task is; but it is a task worth doing. It would be intriguing to explore the paths to which a few scattered conjectures might lead. Let us try the following conjecture.

(I*) There is something about scientists, some aspect of them that is of incomparable significance, with respect to which all scientists are equal, no matter how unequal they may be in talents, achievements, social contributions, and so on.

[20] Ibid., 51.

This conjecture would be prima facie absurd. If their scientific talents and achievements are unequal, from the point of science, what about them could be of incomparable significance? Suppose this, then: The *capacity* or *capability*[21] of a scientist to make a contribution – be it in theorizing, conceiving, and executing a delicate experiment, making observations, and so on – is of incomparable significance. But while that capacity may be accorded some degree of reverence, is respect reserved for the capacity or for what the capacity has *produced*? Suppose, for the sake of argument, it is reserved for the capacity. Then, given wide differences in capacity (in any interesting sense, intrinsic capacity is not a constant, it is just a variable), must not an adequate theory of group rationality – underwriting a scientific image – accord greater honor to the greater capacity?

Let us see where (II*) leads us, given (I*). Suppose we say:

(II*) Even those who are least scientifically talented, or whose scientific achievements are the least, or whose contribution to the sciences is least, are deserving of respect.

This has no semblance to veracity, if read without qualification, and is violated in every which way by what we ordinarily – in science and out – say and do.[22] To every seeker after scientific truth, we owe minimal respect – say, scientific respect – even if after years of labor there is little of that seeker's achievement to report. To every seeker who has discovered a significant scientific truth, we owe considerably more respect – say, achievement respect.[23] For them are reserved Nobel Prizes and Fields Medals. Surely, an adequate theory of group rationality must honor the scientific *deed* – an achievement – more

[21] Had I space, I would have compared and contrasted in detail the views of Putnam and those of his colleague, Amartya Kumar Sen; it would have been a fascinating exercise. Perhaps the reader might do this on his own by juxtaposing the views of Putnam adumbrated here with those of Sen presented in Chapter 2. This much is clear: Assuming the present argument to be sound, whatever plausibility Sen's view has in the context of economic development cannot be simply translated into this context, because capacity and capabilities do not play the same role in the context of science and group rationality as they do in economic development.

[22] There are obviously two senses of "respect" at work here: The respect we owe a humble lab assistant is owed him on moral, Kantian grounds; on the other hand, the respect we owe a Darwin or a Schrodinger is owed him on scientific grounds.

[23] There is an almost an exact parallel in morality, and yet some would deny an analogous distinction there as well; see my paper "Kant: Let Us Compare."

than just the capacity, and accord greater honor to a greater scientific deed. Otherwise, the scientific image it underwrites will be much distorted.

Finally, (III*). It is not clear what plausible counterpart (III) could have. This is the best I could do (keeping in mind what we have so far in (I*) and (II*)):

(III a*) Everyone's happiness (or suffering) in scientific achievement (or failure) is of equal prima facie importance.

Or:

(III b*) Everyone's scientific achievement is of equal prima facie importance.

(III b*) seems to me to be outright false. If, by happenstance, (III a*) turns out to be correct, then the Utilitarian Problem of Group Rationality[24] might well come into its own. To repeat a significant point,[25] if happiness is out of kilter with scientific achievement, then a theory of group rationality that pays deference to that kind of happiness will yield mistaken results. The happiness of the truth seeker must be rooted in the achievement of his science, and that happiness must be proportional to the significance of that achievement. Once again, a theory of group rationality that gives importance to the happiness of the truth seeker must make sure that it allows *science* to play a cardinal role.

Whatever the implications for a society at large, the implications for *how* a group of scientists must organize itself are far from obvious, even granting Putnam's following core assumption (which could serve as a nice general statement of the problem of group rationality):

> The upshot is that *if I am a rational person in the sense of having the aim of making statements which are true humanly speaking, i.e., which can withstand rational criticism now and in the future, then I am committed to the idea of a possible* community *of inquirers.* (In fact, to the idea of a possible community of potentially infinite size, since there can be no such thing as a final inquirer if every inquiry is allowed to be reopened.)[26]

[24] See Chapter 3, section III, 61, and Chapter 4, section I, 75.
[25] See Chapter 4, footnotes 9 and 10, and the text accompanying them.
[26] Putnam, *The Many Faces of Realism*, 54.

Perhaps a society of scientists must minimally make that core assumption, namely, that a truth seeker must be in a society of fellow truth seekers; the scientific proposition he touts must be open to criticism and evaluation by fellow truth seekers in the society; and only in such a society of truth seekers can a scientist genuinely succeed in his aim of making true statements. Let us grant that the core assumption expresses a necessary condition. Yet this will not begin to answer the question "How should a group of scientists rationally organize itself?" Not that Putnam claimed anything more; even so, his answer lacked sufficiency; clearly, it was not an interest in solving the problem of group rationality that moved Putnam here.

Here, then, are three novel questions anyone interested in the problem of group rationality will have to answer. First, granting equality of opportunity, freedom to compete for scarce resources, and the like (the likes of Earnest Everett Just and Barbara MacLintock are treated on a par in this group),[27] by what principle(s) should a group of scientists be structured? Leaving entirely open the question of what defines a structure, the idea behind the structuring is obvious: We want to find a structure that will lead the group of scientists to the common scientific goal(s). Even according respect to every scientist does not ensure that talent does not matter. Consequently, neither the first nor the second principle, neither (I*) nor (II*), is directly helpful. The problem of determining an effective structure for a group of scientists is a remarkably intriguing problem – a problem of first and utter importance – whose connections with social justice we shall trace in the next section.[28]

Second, what common goal(s) of science – such as truth, verisimilitude, or empirical adequacy – must the scientists pursue in light of which they can agree to abide by a common set of principles for structuring their society? If there is no common goal, then there will be

[27] "When relations among scientists become relations of hierarchy and dependence, or when scientists instrumentalize other scientists, again the scientific enterprise suffers." Putnam, *Pragmatism: An Open Question,* 72. In the last century, Earnest Everett Just, a distinguished cellular and developmental biologist, suffered the standard trials and tribulations, as did Barbara MacLintock, whose work on corn eventually won her the Nobel Prize. The one suffered for being an African American, the other for being a woman.

[28] Some aspects of the connection have already been covered in Chapter 2.

fragmentation in the group, a splintering of an objectionable sort such as we encountered in the skeptical view.[29] Note that this is unlike the case of social justice. As we saw Putnam argue, happiness is open to a variety of interpretations, no one interpretation being more reasonable, or ethically more sound, than any other; ergo, one kind of happiness cannot be the end every agent seeks, and so cannot be the one on which a moral philosophy can be founded. Thus follows Putnam's argument for a multiplicity of moral images of the world.

Now, Putnam might be plausibly construed as proposing a multiplicity of scientific images of the world. Kant held that there is exactly one scientific description of the world; Putnam, that there are very many such descriptions. Kant might be deemed a metaphysical realist, Putnam, a metaphysical internalist, one who believes in the possibility of several conceptual systems, ways of Goodmanesque world making – something Donald Davidson famously denied.[30] But if the group of scientists were divided on what image of the world they were making their way toward, the scientists could not engage in a *joint* enterprise, and the threat of fragmentation would immediately follow. Consequently, a common goal will need to be determined, if we are to have a theory of group rationality at all, and discovering that common goal, by any measure, will be a difficult task – certainly difficult on Putnam's view, at any rate, with its insistence on a multiplicity of scientific images of the world.

III. Method, Historical Knowledge, and Reason

Let me now focus on Putnam's next and final lecture, Lecture IV: "Reasonableness as a Fact and as a Value." Putnam's arguments here are too well known to be rehearsed in full, nor is it my task to evaluate them all. Keeping in sight what has just preceded, I want to reconstruct his insights into our historical knowledge and political opinions, show what is significant in Putnam's view of *scientific method*, adumbrate the fascinating Peirce's puzzle he recounts, and delineate the consequences these have for our task of solving the problem of group rationality.

[29] See Chapter 4, this volume, 81–2.

[30] Donald Davidson, "On The Very Idea of a Conceptual Scheme."

Putnam mounts an attack on the claim – he calls it the *argument from non-controversiality* – that factual statements are uncontroversial in a way in which value statements are not. Factual statements can in principle be settled, at least to the satisfaction of all intelligent or educated persons, by observation and experiment. Not so value statements: Weber failed to convince Chinese Mandarins of some Western values. In the long run, disputes over factual statements can be settled; however, those over value statements cannot. But this way of drawing the fact/value distinction, says Putnam, is not even faithful to the hard sciences. For example, opinion on the age of the universe seemed settled at one time, only to be questioned later.[31]

Suppose, says Putnam, we concede that factual statements in the sciences find general consensus or unanimity of opinion; what follows? For one thing, our *historical* knowledge will falter. Our historical knowledge is based on the existence of global causal explanations, or on our knowledge about a counterfactual; yet none of these have been, or can be, scientifically established. Nor would our *political* opinions be safeguarded, because these are dependent on our knowing a variety of consequences that would follow from our plans and actions to stave off, say, outsourcing or polluting the environment. Yet none of these consequences could be established without controversy. However, historical knowledge and political opinion do not lack a truth value for being controversial. In the final analysis, then, it is a matter of taking a stand. "If we find that we must take a certain point of view," says Putnam, "use a certain 'conceptual system', when we are engaged in practical activity, in the widest sense of 'practical activity', then we must not simultaneously advance the claim that it is not really 'the way things are in themselves'. . . . Our lives show that we believe there are more and less warranted beliefs about political contingencies, about historical interpretations, etc."[32]

Consider, next, scientific *method.* As we know, analogies are widely used in both the historical and the physical sciences, and evidence has to be weighed. Putnam offers traditional arguments against Hans Reichenbach's straight-rule of induction, or anything resembling it, and in support of John Stuart Mill's and Nelson Goodman's claim,

[31] Putnam, *The Many Faces of Realism*, 64.
[32] Ibid., 70.

namely, that no merely *formal* criterion can perform the task of assigning proper weights to analogy (what Rudolph Carnap thought to be one of the most intractable of problems in inductive logic). It would follow that "[i]f there is an objective sense in which some inductive inferences by analogy are 'warranted' and others are 'unwarranted', then there must be an objective sense in which some judgments of what is 'reasonable' are better than others, *even if we cannot give a general criterion.*"[33] What is more, "there is no reason to think that a 'method' in this [formal] sense must be independent of the human being's judgments about metaphysics, aesthetic or whatever."[34]

Putnam says that Kant

> does not endorse the kind of conceptual relativity I advocate. ... On the contrary, he thinks that we have exactly one *scientific* version of the world. Yet, I wonder whether there is not a *hint* of conceptual relativity in the fact that each Critique – not just in the first two – we are presented with a different *kind* of reason, and with what might be called a different image of the world to go with each kind of reason: scientific reason, ethical reason, aesthetic reason, juridical reason.[35]

Let me not engage in Kantian exegesis. Yet the question to raise is whether these different kinds of reason are active at the same time on every issue or whether they, on any given issue, operate individually and in isolation from the rest. Thus, is it the tribunal of every kind of reason that sits in judgment on any given issue? Or is it the case that it is the ethical reason that is invoked in discussing ethical issues, scientific reason that is appealed to in evaluating scientific issues, and so on? If the latter is the case, then conceptual relativity does not follow, at least not obviously. For, say, employing solely scientific reason, scientists may find themselves unanimously accepting a covenant on how to rationally structure their group[36] as they would on whether to accept a particular scientific theory, say, the latest theory in cosmology. Or following solely ethical reason, moral agents may find themselves agreeing to Rawlsian principles of justice. However, if the former is the case, reason*s* act conjointly and, as Putnam has argued, method in a formal sense is not

[33] Ibid., 74.
[34] Ibid., 75.
[35] Ibid., 43.
[36] See Chapter 6, 168–71, and Chapter 9, 256–9.

"independent of the human being's judgments about metaphysics, aesthetic or whatever," then the prospect of achieving even a bare majority (let alone unanimity) on any significant matter – say, on the fundamental principles of group rationality or justice – is dim.

IV. Peirce's Puzzle

Finally, Putnam is fascinated by one of Peirce's puzzles.[37] Let me present it, and Putnam's solution to it, and then try to show what bearing it may have on the second problem – the problem of individual rationality – that engages us here. Peirce had a marvelous gift for plumbing the depths of individual puzzles and emerging with powerful insights, even if, says Putnam, he was unable to build a coherent system on these insights or intuitions. One of these puzzles is as follows. Let us suppose that an individual, *S*, is faced with two alternative courses of action, *X* and *Y*. If the probability of success doing *X* is higher than the probability of success doing *Y*, why should *S* do *X*? "We all believe that a rational person [would choose X.] Peirce's question is *Why should he?*"[38]

Peirce at this juncture supposes that *S* is an old man who believes, for whatever reason, that he does not have much longer to live. By his action, he faces two options: eternal felicity or eternal damnation. The issue is moot with respect to what his success frequency would have been had he lived long enough and been involved in a large number of similar circumstances. It is moot because he has not much longer to live. Thus, there will be no further *gambling situation.* Peirce's solution, which Putnam rightly finds unacceptable, is that one can be rational only provided one psychologically identifies with one's community. Putnam cannot be improved upon:

> It is only because I care about what *might* happen to people in similar situations that I do what has the best *chance* in my own situation. My belief that *I* in this one unrepeatable situation am somehow more likely to experience eternal felicity than eternal woe is fundamentally, then, just what Reichenbach said it was, a fictitious transfer, on Peirce's view. What is true, and not fiction or

[37] Putnam, *The Many Faces of Realism,* 80–6.
[38] Ibid., 81.

projection, however, is that my fellows, the members of the community with which I identify, will have eternal felicity... if they follow this strategy.[39]

Putnam immediately queries, "But can it really be that the reason I would choose [to do *X*] is that I am altruistic?" If I do so for *altruistic* reasons, then my reasons are different from what they would be if I had acted in my *own* interest. The issue is not "What ought I to do from an altruistic motive?" It is rather "What ought I to do from a selfish motive?"

Putnam claims that in the face of Peirce's puzzle, any solution to the problem of individual rationality that ultimately derives from the assumption that the ultimate ground for being reasonable is that one will arrive at truth in theory and success in action *more often* if one is reasonable is untenable.[40] What, then, is Putnam's own answer?

In fact, as I came close to the end of my life, and found myself unable to make many more 'bets', then my reasons for doing what is reasonable or expecting what is reasonable should diminish very sharply, on this view. The fact is that we have an *underived*, a *primitive* obligation of some kind to be reasonable, which – contrary to Peirce – is *not* reducible to my expectations about the long run and my interest in the welfare of others or in my own welfare at other times. I *also* believe that it will work better in the long run for people to be reasonable, certainly: but when the question is *Why do you expect that, in this* unrepeatable *case, what is extremely likely to happen will happen?*, here I have to say with Wittgenstein; 'This is where my spade is turned. This is what I do, this is what I say.'[41]

This problem, I want to argue, is more generally applicable – whence its power and fascination. Let us suppose that a scientist, *S*, is deciding which research path to follow. There are two paths, *X* and *Y*. If *S* acts altruistically and does *X*, which has a greater chance of succeeding – in other words, acts to ensure that his scientific *group* will benefit in the long run – then he has *a* reason for acting. But this fails to be a reason why this choice is rational for *him*. Such circumstances, in which he has to make substantial research choices, even if not exactly unrepeatable, will be far too few. Perhaps he begins his career as a scientist with several options, and as time passes by, his options seriously diminish.

[39] Ibid., 83.
[40] Ibid., 84.
[41] Ibid., 84–5.

Therefore, at the start, he can act in a rational self-interested manner; over time, his rationality increasingly shifts to what is good for his community. Should the scientist, making a choice that is rational from *his* point of view, end up saying, "Here is where my spade is turned?"

Let us go a step further. Regard the entire group of scientists as an individual and say that *it* is confronted with a huge choice – opt for Ptolemy or Copernicus? Opt for creationism or the theory of evolution and natural selection? Opt for the Copenhagen interpretation of quantum mechanics or not? Opt for the Genome Project or not? – What ought *it* to do? It is clear that a Peirce-like argument would apply in such a case as well. Will group decisions, then, in such nonrepeatable circumstances be just as irrational as the decision of Peirce's old man?

Putnam does not think that this will lead to Feyerabend-style anarchism,[42] but it is difficult to see why not.[43] Each scientist belonging to a distinct subgroup might make a unique decision, analogous to Peirce's old man's, and be no less irrational for it. Secondly, Putnam, without argument, presupposes *what* it is to be reasonable and then takes it as an argument for the demand to be reasonable. But these are separate issues. First, there is the question "Why should I be reasonable?" or "Do I have an obligation to be reasonable?" Yes, says Putnam, we do: "The fact is that we have an *underived*, a *primitive* obligation of some kind to be reasonable." This is a rather important question to which Putnam has drawn our attention, and which I have not seen asked. It is analogous to the question "Why should I be moral?"

The second question is, "Given that I have this primitive obligation to be reasonable, wherein lies my acting reasonably?" I fear Putnam took the answer to the first question as an answer to the second. Usually, an answer to the question "Why should I be moral?" does not answer the question "Given that I ought to act morally, how shall I act?" The two questions in ethics are not thought to be the same question, nor is the answer to one thought to answer the second. Similarly, the two questions of rationality – "Why should I be rational?" and "Given that I

[42] Ibid., 86. See also Putnam, "Philosophers and Human Understanding," in his *Philosophical Papers*, Volume 3, 192–7.

[43] Comparing Putnam and, in particular, Feyerabend's "Democracy, Elitism, and Scientific Method," in his *Knowledge, Science, and Relativism*, may not be an entirely fruitless exercise.

ought to act rationally, how shall I act?" – are not the same. Analogously to ethics, the first question probes deeper than the second question; one might say that the answer to the second question presupposes an answer to the first.

Now, answering the first fundamental question – "Why should I be rational?" – cannot simply be a matter of how my spade is turned. Otherwise, two scientists turning their spades in different directions can each claim to be acting reasonably. Not only that, we would have a *very* curious result. If each scientist were to act in his own best interest – as defined by the way *his* spade is turned – we would have a group of scientists each acting rationally, but with no way of determining the rationality of the group. On the other hand, if each scientist were to act in the best interest of his scientific group, we would have a rational group, but one that could not be founded upon the interest or rationality of the individual. In such an event, *How would a scientist acquire allegiance to his group?* As Bernard Williams might have said, "Why should *he* be listening?"[44]

This bears closer examination. Ultimately, for Putnam, the matter rests in Wittgenstein. Here, then, is *Philosophical Investigations*, section 211:

> How can he *know* how he is to continue a pattern by himself – whatever instruction you give him? – Well, how do I know? – If that means "Have I reasons?" the answer is: my reasons will soon give out. And then I shall act, without reasons.[45]

Then consider section 217:

> "How am I able to obey a rule?" – if this is not a question about causes, then it is about the justification for my following the rule in the way I do.
>
> If I have exhausted the justifications I have reached bedrock, and my spade is turned. Then I am inclined to say: "This is simply what I do."[46]

One might construe Wittgenstein – "I have reached bedrock, and my spade is turned"– variously thus:

[1] This is where *my* spade turns.
[2] This is where *our* spade turns.

[44] See Chapter 1, section III, 18–20.
[45] Ludwig Wittgenstein, *Philosophical Investigations*, 84e.
[46] Ibid., 85e.

[3] This is where *human nature*'s spade turns.
[4] This is where *rational nature's* spade turns.

Putnam's remark in a footnote[47] reveals where he stands with respect to alternatives [1] through [4]. He says, "That Wittgenstein here uses the first person – where my spade is turned – is very important; yet many interpreters try to see his philosophy as one of simple deference to some 'form of life' determined by a community."[48] And, I think, it might be interesting to see how the other alternatives are eliminated – not that Putnam himself proceeds in any such manner – and then to examine what is left for plausibility. Proceeding backwards, clearly [4] must be eliminated, for Putnam is no Kantian in one significant respect. As he says, "the way in which I shall diverge the most from Kant is in not trying (or even pretending to try) to derive ethics from reason alone."[49] In saying Putnam is un-Kantian in one respect, I am guilty of a generalization. If we assume – correctly, I believe – that a theory of group rationality is a normative theory, like an ethical theory, then if ethics cannot be derived from reason alone, then the prospect of deriving a theory of group rationality from reason alone, by Putnam's lights, looks just as dim. Consequently, Putnam cannot be seen as translating Wittgenstein's dictum into [4].

Just as clearly, [3] must be eliminated, because Putnam is no Humean either. As Putnam says, "The whole Kantian strategy, on this reading at least, is to *celebrate* the loss of essences, without turning back to Humean empiricism."[50] Putnam could not, therefore, commend a Humean solution to the problem of group rationality for reasons analogous to the ones he gives in the case of ethical theory. And if we include, as a variant on [3], essence or *ergon*, in talking about humans as did the medievalist, the answer, in view of the foregoing, would not be any different, and Putnam would reject that variant as well. Might

[47] Putnam, *The Many Faces of Realism*, 91, footnote 32.

[48] Putnam also refers to Stanley Cavell, *The Claim of Reason*, especially Chapter 5, "The Natural and the Conventional." However, Putnam once also conjectured that for Wittgenstein it was some subset of our institutionalized scientific norms (forms of life?) that determines what is right and what is wrong, and that there is nothing more to right and wrong beyond this language game of science; see Putnam, "Philosophers and Human Understanding," in his *Philosophical Papers*, Volume 3, 186–7.

[49] Putnam, *The Many Faces of Realism*, 56.

[50] Ibid., 52.

Putnam opt for [2], reading *our* as short for *those who share a common form of life,* and think that this might be helpful in taking us toward the solution to the problem of group rationality? In this reading of [2], we are enjoined to look for parallels between what constitutes *a common form of life of the scientific group* and what constitutes a Wittgensteinian *common form of life.* But in view of Putnam's footnote cited earlier, that too would surely be an unpalatable option for Putnam. Thus, we are left with [1].

What puzzles me about [1] is that it sits rather ill with Putnam's other texts, texts that cry out for a distinction. The distinction I have in mind is one that Putnam makes not only in the context of ethics (where it *may* be more plausible), but in the context of science as well (where it *is* arguable). Thus, we have *individualistic decisions* and *universalistic decisions,* both in ethics and in science. Consider first the case in ethics, and the individualistic decision. Pierre in Jean-Paul Sartre's *Existentialism and Humanism* has to decide whether to stay at home and nurse his ailing mother or to abandon her and join the resistance. Not exactly pleasant alternatives: Either be a good son, but fail in the defense of one's nation; or be a good citizen, but an uncaring son. This is not a case of justifying a social policy; one cannot invoke consequences to show justification. Pierre, says Putnam, is interested in doing *right,* and Dewey's consequentialist views are of no help here. "One of the things that is at stake in Pierre's situation is his need to decide who Pierre *is.*"[51]

Putnam, with his entirely justified lament elsewhere, provides us with an answer to the foregoing:

> Nor should commitment to ethical objectivity be confused with what is a very different matter, commitment to ethical or moral authoritarianism. It is perhaps this confusion that has led one outstanding philosopher[52] to espouse what he himself regards as a limited version of "non-cognitivism", and to say "Concerning what 'living most fully' is for each man, the final authority must be the man himself."[53]

This allows us an opening, and we can respond to Putnam thus: If the final authority, concerning what is "living most ethically," cannot be the

[51] Putnam, *Renewing Philosophy,* 191.
[52] David Wiggins in "Truth, Invention, and the Meaning of Life."
[53] Putnam, *Reason, Truth, and History,* 148–9.

moral agent, then the moral agent cannot account for his rationality by merely ending with "This is where *my* spade is turned" or "This is what *I* have decided to be."[54]

Consider next the case in ethics, and the universalistic decision. Dewey is enormously interested in human flourishing, although he does not think that knowledge of that flourishing can be had on an a priori basis, through consulting our reason alone, in isolation from practice and experience. He relies "on our capacity intelligently to initiate action, to talk, and to experiment. Dewey's justification is not only a social justification – that is, one which is addressed to *us* as opposed to being addressed to each 'me' – it is also, as I said at the outset, an *epistemological* justification."[55] Now, William James believed "that Pierre has the right to believe and act 'in advance of the evidence'."[56] And just why does James think that? The answer is given in purely pragmatic terms: "James would say that what these cases have in common is that it is valuable, not just from the point of view of the individual, but from the point of view of the public, that there should be individuals who makes such choices."[57] Or: "[O]ur best energies cannot be set free unless we are willing to make the sort of existential commitment

[54] Putnam then goes on to say that even in the religious context, the answer is the same. Citing the Jewish tradition, Putnam says that Rabbi Susiah said that in the hereafter, God is not going to ask him if he had been Abraham, Moses, or Hillel, but rather "Have you been Susiah?" Putnam claims that in a similar manner, Pierre wants to be Pierre, to "become" – as Kierkegaard might have said – "who he already is." See also Putnam, *Renewing Philosophy*, 194. I find the Kierkegaardian view unpalatable, but I will not rehearse that argument here; an interested reader might consult, my *The Toils of Understanding*.

Let me return to Pierre, then. The problem is contained in Putnam's formulation of the question. I should think that Pierre above all wants to be *moral*. As he anguishes over the decision, he wants a truthful moral answer he can honor in his deed, and *thereby* to regard himself as being true to himself. If questioned, he can defend himself; if others criticize him rightfully, he acknowledges the error of his ways. But these wants and acknowledgments presuppose some standards by which one's deeds are judged. Otherwise, one can simply get away each time by saying, "Here is where I stand." Now, the odd thing is that Putnam himself adopts this critical stance. Otherwise I cannot understand his criticism of Bernard Williams; see Putnam, *Renewing Philosophy*, 181.

[55] Ibid., 182.

[56] Ibid., 192. This surely needs defense, because the issue was not whether Pierre has a *right* to believe and act in the manner he chooses to (given all the correct constraints), but whether what he chose to do was *morally* right. The query was whether his mere *choosing* to do something made the act in and of itself morally right.

[57] Ibid., 194.

that this example illustrates."[58] Or, and perhaps this clinches it: "At the same time, James never fails to see the need for a *check* on existential commitment."[59]

What does a check on existential commitment accomplish? For example, if a sufficient number of moral agents were to make individualistic decisions that harmed the society in the long run, would that mean that their respective decisions were not right? On the other hand, if such decisions proved beneficial to the society in the long run, would that mean that these decisions were right? If the pragmatist answers these questions in the affirmative (as indeed he must), it would dissolve what Putnam thought to be of inherent value in the existential commitment or decision itself. Ultimately, it would follow, existential commitment is answerable to the empirical check.

Now, consider the case of science, and the individualistic decision. Putnam recounts a story told to him by Gerald Holton. Max Planck, when pushed, and pushed hard, by the Berlin physicists to give them at least some experimental evidence for preferring Einstein's theory of special relativity, could supply none, except to say, "It's simply more sympatico to me."[60] And yet Planck had influenced elite physicists to consider that theory seriously.[61] Group rationality might throw some light on this. One might say that so long as other scientists are successfully propelling the work of the group, the maverick behavior of *some* scientists might be tolerated (within limits). If too many of them turn

[58] Ibid.

[59] Ibid., 195.

[60] Ibid., 193. If Planck was persuaded of the correctness of the early quantum theory despite the lack of experimental evidence, Freeman Dyson offers a sterling example of a theory left unacknowledged by some despite its incredible successes. This time the younger physicists, like Julian Schwinger, Richard Feynman, and Dyson in America and Sin-Itiro Tomonaga in Japan, are conservatives, and the older generation of physicists, like Albert Einstein, Paul Dirac, Werner Heisenberg, Max Born, and Erwin Schrodinger, are the would-be revolutionaries. The efforts of the younger physicists, built on the initial work of Dirac, ushered in the modern theory of quantum electrodynamics. Experiments were performed to test the theory, and they corroborated the theory to eleven decimal places. After that success, Dyson ventured to ask Dirac if he was pleased with the theory that he, Dirac, had created twenty-five years earlier. Dirac, as was usual with him, was silent for a while, and then said: "I might have thought that the new ideas were correct if they had been not so ugly." "The World on a String," 16–17.

[61] See the interesting paper by Alexander Reuger, "Risk and Diversification in Theory Choice."

maverick, the group's work as a whole suffers, and justification of such deviant behavior is hard to come by. What gives voice to our objection here are Peirce's arguments against the Method of Authority.

Methods, Putnam has argued, are not algorithmic devices; in order to use them one has to weigh evidence, and weighing will involve values. Consequently, it is worth asking that if a nonalgorithmic method involves maxims[62] whose epistemic worth is best measured by putting it into practice and allowing it to be freely criticized by the members of the group to which one belongs, then how is that view to be reconciled with what Putnam tells us here with respect to individualistic decisions, both in and out of science? Putnam asserts that James's claim was "that science would not progress if we insisted that scientists never believe or defend theories except on sufficient evidence." However, "[w]hen it comes to institutional decision, . . . then it is important that we apply the scientific method."[63] I take Planck's decision to be individualistic, and the institutional decision to be the universalistic decision. We are inevitably led to the conclusion that as in ethics, so in science: If individualistic (or existential) decisions are ultimately answerable to universalistic decisions whose value in turn is decided by their practical worth, individualistic decisions – as markedly distinct ways of making a choice – seems to have no genuinely autonomous function at all.

Let me sum up this section: The problem of group rationality would be rather difficult to solve if each individual scientist were to take a stand and say, "Here is where my methodological spade turns, and I know no other." The specter of relativism would loom large on the horizon.

V. Ultimately, Relativism?

With a view to just how such a charge of unbridled relativism might go, consider what follows. Transcribing what Putnam learned (and approved) from C. West Churchman:[64]

[62] A maxim offers fairly general guidance.

[63] Putnam, *Renewing Philosophy*, 193.

[64] Who received it from his teacher, A. E. Singer, who in turn might have learnt it from *his* teacher, William James; see, Putnam, *Pragmatism: An Open Question*, 13–14. The argument that follows considerably overlaps with the final chapter of Putnam's *The Collapse of the Fact/Value Dichotomy and Other Essays*. As we proceed with the argument

(1) Knowledge of facts presupposes knowledge of theories.
(2) Knowledge of theories presupposes knowledge of facts.[65]
(3) Knowledge of facts presupposes knowledge of values.
(4) Knowledge of values presupposes knowledge of facts.[66]
(5) Knowledge of facts presupposes knowledge of interpretations.
(6) Knowledge of interpretations presupposes knowledge of facts.[67]

Add what we adverted to earlier, namely, that there is scientific reason, ethical reason, aesthetic reason, and juridical reason (and these are neither reducible nor translatable to any other),[68] and we would find that these – facts, theories, and interpretations – would substantially vary depending, say, on the scientific, ethical, or aesthetic values that a scientist holds. Among the scientific values are coherence, plausibility, reasonableness, simplicity, and – I should add one value not on Putnam's list – testability or falsifiability.[69] Lest one is minded to separate ethical values from scientific values, Putnam forewarns: "[T]he theory of inquiry, or the theory of the admirable in the way of scientific conduct, presupposes ethics."[70]

(3) is ambiguous: it would have us assert that either (*a*) their knowledge of the fact is the same, but the two scientists evaluate the fact differently, giving the same fact different weights, or (*b*) they in fact acknowledge different facts, given their different values. (*a*) does not comport well with Putnam's holism, while (*b*) leads straightforwardly to relativism.

Suppose there are two scientists, S_1 and S_2, who hold different values, V_1 and V_2, respectively; indeed, let us even suppose that their ethical values are the same. Now, in evaluating, say, the rival theories of Einstein's theory of gravitation and Alfred North Whitehead's 1922

here, and to minimize repetition, I beg the reader to keep in mind Chapter 4, sections IV–VI, this volume, wherein we had the skeptic's view. This will facilitate my making plausible the claim that Putnam's view and the skeptic's view are hard to distinguish.

[65] To demonstrate that no sharp boundary line can be drawn between fact and theory, see Putnam's arguments against Ian Hacking on the reality of positrons: *Pragmatism: An Open Question*, 60–1.

[66] Ibid., 14.

[67] Ibid., 18.

[68] Following Kant: ibid., especially 29, 30–1.

[69] But it is on Quine's list, as Virtue IV, called refutability (*The Web of Belief*, 50), no less than – of course – on Popper's.

[70] Putnam, *The Collapse of the Fact/Value Dichotomy and Other Essays*, 135.

theory, S_1 espouses the values of coherence and simplicity, V_1, while S_2 is enamored by testability and beauty, V_2. Both theories squared with special theory relativity, both predicted light deflection by gravitation, and both explained the perihelion of Mercury as well as the exact orbit of the moon.[71] "To borrow and adapt Quine's vivid image, if a theory may be black with fact and white with convention, it might well... be red with values."[72] Let us grant that not only (1) and (2), but (5) and (6) as well, are innocuous. But (3) would assert that S_1's knowledge of the facts, given his upholding of value V_1 (coherence and simplicity) would be different from S_2's knowledge of the facts, given his value, V_2 (testability and beauty).

There is a potential dilemma: Either S_1 and S_2 both understand and evaluate the *same* theories, namely, Einstein's theory of gravitation and Alfred North Whitehead's 1922 theory, in which case the strand of values *can* be untangled from the web; Putnam's holism would be

[71] Ibid., 142. Putnam has claimed that this example demonstrates the mistaken idea that "when two theories conflict, scientists wait until the observational data decide between them, as Popperian philosophy of science demands they should." It is mistaken because the scientific group did not wait for any such observational data. Indeed, fifty years before any such observation was available, the group had abandoned Whitehead's theory.

How plausible is the Putnamian perspective? First, just what exactly does this example show? Does it show that falsifiability, contrary to both Popper and Quine, does not play any significant role in science? No, that would be patently absurd, nor does Putnam maintain it.

Second, assuming, then, that falsifiability plays a role in the decision-making process of science, Putnam's argument must be that in *this* case it played no significant role and that, more generally, whether the principle of falsifiability should play a role in the decision-making process must be judged on a case-by-case basis. This leaves us with the task of specifying the conditions under which falsifiability is not operative.

Third, would similar claims be true of *all* values, and not just the value of falsifiability? And if not, why not?

Fourth, would the decision of the group of scientists be what ultimately justifies the correctness of a decision? No, for Putnam himself has provided us with powerful arguments to the contrary. If Putnam does not want to settle for individual psychology, as Quine has advocated, he would find the proposal to settle for group psychology no less unattractive. So, then, why is it relevant to have pointed out what the group of physicists did decide in the case of Einstein versus Whitehead? What ultimately justified that decision?

Fifth and finally, how would Putnam justify the decisions of particular scientists (see footnote 60) to accept a theory precisely because it has successfully met the Popperian criteria of severe testing (Dyson) while others reject it on grounds of beauty (Dirac)?

[72] Vivian Walsh, "Philosophy and Economics," cited in Putnam, *Pragmatism: An Open Question*, 18.

in need of a qualification. Or the content of the two theories cannot be disentangled from the values that S_1 and S_2 bring to bear on them, in which case they, S_1 and S_2, are not evaluating the same theories. (Please note: We have assumed that the ethical values that S_1 and S_2 bring to bear are constant; the problem for Putnam's holism would only be compounded if we allowed for variations there as well.)

Putnam expands on (3), crucial for our purposes, as we have seen, as follows:

> This is the position I defend. It might be broken into two separate claims: (i) that the activity of justifying factual claims presupposes value judgments, and (ii) that we must regard those value judgments as capable of being *right* (as "objective" in philosophical jargon), if we are not to fall into subjectivism with respect to the factual claims themselves.[73]

This leads to an even deeper problem. Let us agree not only with (i) but also with (ii): First, when S_1 and S_2 are engaged in the activity of justifying the acceptance of a theory – either Einstein's theory or Whitehead's – they must presuppose value judgments. Second, value judgments must be right and objective, if subjectivism is to be avoided. How, then, does one know which values, and hence which value judgments, are right and objective? One cannot say: "Look at the *practice* of science. When certain values have been espoused in the history of science, they have led to successful results or decisions; these values have led to scientific theories that were verisimilar or nearer to the truth than the theories they succeeded." One cannot say that because such claims will, in turn, have to be tested. Now, one can test such claims neither by the Quninean route, nor by the Reichenbachian route, nor by the Popperian route – routes anathema to Putnam.[74] Testing them by invoking the very values under test would be uninterestingly circular. One might opt for psychology, as did Quine. But that, understandably, Putnam finds unpalatable.[75] One cannot find in empirical matters answers to normative questions.[76] There is no fourth alternative I can envision. (I cannot forbear to remark that if ethical values are as objective as scientific values, then it is not clear what happens to Putnam's ethical plurality of values; surely, conjoining ethical plurality

[73] Putnam, *The Collapse of the Fact/Value Dichotomy and Other Essays,* 137.
[74] Ibid., 137–42.
[75] Ibid., 139.
[76] See Sarkar, *A Theory of Method,* 17–26.

of values with plurality of scientific values, values by which to judge the credentials of a scientific theory, would lead ineluctably to relativism.)

Even deeper still, to get at that relativism would be to deny, with Wittgenstein, that there is any genuine problem of knowledge – knowledge of facts, values, and theories. For knowledge has no nature, and hence no possibility of laying down in advance what will be deemed rational or legitimate criteria of knowledge; the use of *know*, therefore, will always be in a state of Heraclitean flux.[77]

The intentional is ubiquitous, says Putnam. The idea emerges most powerfully in Putnam's criticism of Jerry Fodor's attempt to reduce the intentional to the nonintentional (as described by the special sciences, such as geology, evolutionary biology, and psychology). As both philosophers recognize, these sciences – indeed, any science – invoke ceteris paribus clauses, laws, causation, and counterfactuals. Consider the counterfactual. In Putnam's theory, there are *strict counterfactuals* (those that are derivable from the laws of nature) and then there are the nonstrict kinds. The latter are often invoked to give confirmation to a law even though there is a possible world in which the antecedent of the counterfactual is true and its consequent false. But then we have to *judge*, based on our *interests*, that such a counterfactual is no evidence against the law whose confirmation we are investigating. Such a judgment is eminently plausible, thinks Putnam – indeed, it is a judgment that rests on a *rational* point of view.[78]

But this poses a problem for Putnam. On the one hand, if there is a rational point of view, then no matter how different one's actual interests, there are some interests that you must perforce objectively have (in that context), and judged rationally from this perspective of objective interests there is only *one* answer forthcoming about the confirmability of the law in question. "What it is right to say in a given context cannot always be established to everyone's satisfaction; but it is nonetheless the right thing to say."[79] Ultimately, therefore, there can be only *one* picture of science, albeit from a particular, rational point

[77] Putnam, *Pragmatism: An Open Question*, 32. My attempt, in part, has been to show that when Putnam, arguing against Richard Rorty, later defends the notion of one language game (or one criterion of knowledge) as more defensible than another, he has divested himself of premises that could support his claims; see ibid., 37–8.

[78] "In my opinion, counterfactual conditionals and causal statements presuppose what I call 'the point of view of reason,' " Putnam, *Renewing Philosophy*, 61; see also 55.

[79] Ibid., 77.

of view. Putnam offers no such theory of rationality. Note, too, that on this picture the analogy between scientific image and moral image breaks down: A society may be adorned with multiple moral images, by not with multiple scientific images.

On the other hand, if there is a rational point of view and what is rational depends on one's actual interests, then given the multiplicity of interests, there will be a multiplicity of rational answers. "But what makes some interests more reasonable than others? The answer is that reasonableness depends on different things in different contexts."[80] Some, given their interests, will regard a law as confirmed by a counterfactual; others, given their interests, will be less sanguine. There will be, in principle, no possibility of an answer about the confirmability of the law. Consequently, it would be difficult to see how the following would hold up: "What we say about the world reflects our conceptual choices and our interests, but its truth and falsity is not simply determined by our conceptual choices and our interests."[81]

One might plead that the foregoing is guilty of ignoring the normative, an inescapable element in rendering such judgments: "Yet the explicit recognition that language games are human activities in which what is right and wrong is not simply conventional, is not simply determined by consensus, but is something that requires evaluation, is troubling to many a contemporary sensibility. . . . the rightness of what is said – *is* a normative notion."[82] But much more would have to be granted to Putnam in order to escape the charge of relativity; we would have to grant not only that there is a normative element in the kind of judgments we are talking about, but also that there cannot ultimately be several competing norms. For Putnam says,

> Truth conditions are fixed by conditions of warranted assertability together with conditions of the form "epistemic conditions A are better than epistemic conditions B for making judgments of warranted assertability about S." Note that these latter conditions are also normative. Of course, a metaphysical realist might hold that (what we take to be) conditions of warranted assertability do not constrain truth conditions *at all*; but I do not believe that this can be a position [Bernard] Williams would find congenial.[83]

[80] Ibid., 66.
[81] Ibid., 58.
[82] Ibid., 77.
[83] Ibid., 214, footnote 32.

Let me not argue in defense of the claim of the metaphysical realist that conditions of warranted assertability do not constrain truth conditions *at all*; rather, let me try to draw out the consequences of Putnam's views. There cannot be competing normative, epistemic criteria. For suppose there were. Then one set of criteria may proclaim sentence *S* to be true, while another set of epistemic criteria may find it to be false. If there were competing sets of such criteria, they would have to be rank ordered in the way in which Putnam does in the last quoted sentence. (Might the skeptic ask, "Why would such a criterion for evaluating competing epistemic criteria itself have no competitors?") Would these epistemic norms themselves not be guided by interests, and would we not also have to presuppose a uniformity of these interests from a rational point of view?

What we have, then, is a Quinean web of facts, values, theories, and interpretations, so intermingled as to produce a coloration distinct from that produced by a web of *other* facts, values, theories, and interpretations. Change one ingredient – say, values – and it will infect the remaining ingredients, thus producing a different coloration. One might safely conclude that two scientists with different values have incommensurable positions and that there does not seem, in principle, a way of negotiating between them.[84] Feyerabend would have been pleased.

VI. What Lies at Bedrock

However, what saves us from this treacherous conclusion is Putnam's powerful insight encased in his remark quoted earlier: "The fact is that we have an *underived*, a *primitive* obligation of some kind to be reasonable." Let us refer to this as the Putnamian idea. There is no space in this work to delve into this fecund idea, save for some scanty, programmatic concluding remarks.

One might think this Putnamian idea to be so thin that nothing substantial could be made to stand on it. I see the idea differently – as something just big enough to build our epistemological house on. Suppose we said that the Rawlsian notion of the original position might be regarded as ultimately dependent on the Putnamian idea. Rawls

[84] On the notion of negotiation, see Chapter 6, section IV, this volume.

placed moral agents in the original position, giving them access to precious little in that position in the hope of showing that moral agents can reason themselves to a consensus about liberty and how resources in the community are to be distributed. Must this not presuppose, at least as a necessary condition, the Putnamian idea that moral agents in the original position have an *underived, primitive* obligation of some kind to be reasonable?[85] To this slim idea Rawls added a great deal – a master stroke, if there ever was one, in the history of philosophy – to make it deep, powerful, distinctive.

There is a lesson in this for us. A philosopher interested in solving the problem of group rationality might invent a notion parallel to the Rawlsian notion, taking the Putnamian idea as his point of departure.[86] What brushes off relativism from Putnam's view is the claim that the primitive, underived obligation to be reasonable is the *same* for all scientists,[87] and with respect to *this* obligation they are governed by the *same* general rules.[88]

[85] When Davidson says that his "approach looks at language from the start as a social transaction and therefore concentrates on what one person can learn about another in the context of a shared world" ("Epistemology and Truth," 190), his learner must have, among other things, minimal resources of reason. It is this minimum, I contend, in conjunction with the Putnamian idea that the scientific agents in the original position have an underived obligation of some kind to be reasonable, that is objective and sufficient to produce the kind of core agreement between scientists on how to rationally structure their group.

[86] I am trying to attempt this in the next volume; but see Chapter 6, section IV.

[87] "What I am saying is that the 'standards' accepted by a culture or a subculture, either explicitly or implicitly, cannot *define* what reason is, even in context, because they *presuppose* reason (reasonableness) for their interpretation. On the one hand, there is no notion of reasonableness at all *without* cultures, practices, procedures; on the other hand, the cultures, practices, procedures we inherit are not an algorithm to be slavishly followed." Putnam, "Why Reason Can't Be Naturalized," in his *Philosophical Papers*, Volume 3, 234.

[88] The idea of primitive, underived obligation has remarkable affinity with the Kantian notion of *conscience*: "[S]o too, conscience is not something that can be acquired, and we have no duty to provide ourselves with one; rather, every human being, as a moral being, *has* a conscience within him originally." Kant, *The Metaphysics of Morals*, 160; see especially 160–1. Just as one cannot have a duty to acquire a conscience if one does not already have it, so also one cannot have a duty to acquire a primitive, underived obligation to be reasonable if one is not already reasonable. In view of this, Bernard Williams is surely amiss in thinking that Kant was engaged in the task of deriving a justification for moral behavior from purely nonmoral considerations; see Williams, *Ethics and the Limits of Philosophy*, Chapter 4.

Or look at it thus. Putnam has insistently maintained that the notion of truth shorn of any connection with an epistemology is a vacuous notion. Says Putnam, "[T]ruth itself, on my view, is an idealization of rational acceptability. The idea that we have some notion of truth which is totally independent of our idea of rational acceptability just seems untenable to me."[89] The notion of an idealization of rational acceptability protects truth from the following obvious objection, namely, that if truth were harnessed simply to a particular method or an epistemology, since there is a proliferation of methods in the history of science, each scientific group with its own distinctive method would be a peddler of its own version of truth. There would then be as many notions of truth as there were methods; this would lead to crass relativism. Now, Putnam says very little on the precise nature of this idealization of rational acceptability with which truth is connected; nor shall I now speculate as to Putnam's answer. What I wish to do is to make a conjecture, in this concluding sentence: If the notion of an *idealization of rational acceptability* (whatever *it* is)[90] can be harnessed to what is *reasonable* (whatever *that* is),[91] as in "The fact is that we have an *underived*, a *primitive* obligation of some kind to be reasonable," Putnam might show us a most beguiling way out of the impasse relativism threatens.

The general rules that are emphasized are sufficiently slim to embrace a variety of scientific views or conceptual relativities. It is predicated on the venerable view that learning or using any concept – in this case, the concepts involved in the Putnamian idea – must be a matter of rule-governed activity; see R. M. Hare, *The Language of Morals*, 60.

[89] Putnam, "Philosophers and Human Understanding," in his *Philosophical Papers*, Volume 3, 200; see also 210, 231. Putnam has taken this stance since *Meaning and the Moral Sciences*.

[90] Not to muddy the waters too much, it would be useful to keep in mind the following caveat: "[B]ut we cannot reasonably expect that *all* determined researchers are destined to converge to one moral theory or one conception of rationality." Putnam, "Philosophers and Human Understanding," in his *Philosophical Papers*, Volume 3, 202. Does this mean that Putnam allows for the possibility of several incompatible conceptions of rationality to be, idealizations of rational acceptability? If so, my conjecture is doomed, and relativity would threaten us again.

[91] It is worth noting, though, that Putnam is not observing the Humean distinction between *rational* and *reasonable*; see ibid., 186. Observe, too, the following: "It is true that some judgments of 'reasonableness' must be made simply on the basis of ultimate intuition; not everything can be proved. And it is true that here responsible and careful thinkers may disagree, and may even consider that the other accepts illegitimate arguments. But this is very far from saying that our attempts to be rational are a fraud." Ibid., 203–4.

9

The Nine Problems

There were several different ways of bringing this book to a close. One way was to outline the reasons for accepting one of the theories of group rationality herein examined as unqualifiedly the best; but there is no such theory, for there are no such reasons. Another way was to demonstrate one of the theories as the least objectionable, but that task – and I am not at all sure that it is an unworthy task – left me without motivation. A third way was to list the virtues and vices of each theory and show how a concatenation of these virtues, without the attendant vices, can be welded into a single theory of group rationality; but it was doubtful if that could be accomplished in a single chapter. A fourth way was to highlight the key arguments of the book, but I saw little value in repeating them. Ultimately, I decided to present the cardinal problems of group rationality – nine in all – as a substantial body, not merely a filigree, and to color it from a distinctive point of view. Shaped by the arguments thus far, such a view might assist in our future inquiry into the subject of group rationality, if only as a view to argue against.

The skeptic in philosophy has standardly played the role of one who demands justification even when we are at the outermost limit of our philosophical investigations. Thus, in presenting a list of the fundamental problems in group rationality, I have aimed at two things: First, in several places I have tried to explore and state – clearly, if schematically – what each of the problems seeks by way of a solution under the pressure of the skeptic's demands. Second, I have tried to show how these basic problems are connected: Some may see in this an unveiling of a unified structure – ordering, ranking – of these problems; others may discern only a web. But not the least interesting are the assumptions underpinning these problems, for these unearthed assumptions may have their own role to play in answering the skeptic.

In section I, I present the first three problems. I begin by briefly restating the problem of group rationality; a careful formulation and analysis of it is unquestionably our primary task. The second problem is the problem of deciding what the common aim of science should be around which the group can center, coalesce; and the third problem turns on defining the basic structure of the society of scientists. Section II presents the next three problems: The fourth problem is one concerned with delineating what I call the Council of Scientists, somewhat analogous to the Rawlsian original position; the fifth problem is about the type of reasoning that must be involved in the Council, individual rationality again taking the limelight, this time against the backdrop of the basic structure; and the sixth problem deals with how allegiance is engendered and sustained among scientists. Finally, in section III, I present the last three problems: The seventh problem is about the ultimate principle of rationality – one might call it the universal law of rationality; the eighth problem pertains to the ultimate worth of science; and the ninth and last problem is about utopias.

No one will be blamed – surely – for declaring that the sequence of arguments in this chapter is but a series of footnotes to Rawls.

I. The Problem, the Common Aim of Science, and the Basic Structure

Our *first* problem is, of course, the proper analysis of the problem of group rationality. Having outlined a variety of problems under that rubric, we settled upon the following version (and the major theories of group rationality examined in this book were assessed in terms of how adequately they answered this version of the problem): Engaged in a joint enterprise to pursue a set of shared (partially overlapping?) goals, what ought to be the structure of a society of scientists – where structure is defined in terms of method – at a given time, or over a period of time, that would enable that society to reach its goals better than it would under any other structure?

Each of the elements in this statement of the problem – goals, structure, and method – will be taken up shortly, and analyzed, but for now we must redraw a significant distinction. This is the distinction between the static problem and the dynamic problem of group rationality.[1] The static problem of group rationality (our sole focus thus far) is to determine whether a group of scientists, at a given time, is structured rationally; the dynamic problem is to determine whether a group of scientists has evolved to a rational structure or to a more rational structure. As will become evident, these problems are closely linked not only to the new problem of demarcation but also to a theory of method (or meta-methodology).

The new problem of demarcation asks: "How is a good method to be distinguished from one that is not?" or "How is a good method to be distinguished from a pseudo-method?"[2] One might say that the new problem of demarcation is one of the problems that a theory of method must address. Now, if the rationality of a group is dependent upon a method (on which it is structured), and if the method must be viable, then the new problem of demarcation is indispensable to the problem of group rationality. The solution to the new problem of demarcation will tell us the conditions under which a method is viable. If we are to speak of a group of scientists having become more rational, then a theory of method is indispensable, because a part of what a theory of method deals with is whether there has been growth or improvement in method.[3] If growth or improvement in method has in fact occurred, then a society founded upon an improved method is more rational; not otherwise. While the skeptic can undermine or challenge a purported solution to either of the two problems, there is no discernible argument he can give that would undermine the distinction between the static and dynamic problems of group rationality (unless it be in the guise that the only solution to *any* problem of group rationality is simply "Anything goes," and that therefore the distinction is not worth making).

Our *second* problem is to show *how* scientists can come to agree upon a common aim of science, an aim that will gather them together into a

[1] See Chapter 1, 9, and Chapter 6, 160, this volume.
[2] See Chapter 7, section V.
[3] See Sarkar, *A Theory of Method.*

single group, enlisting their loyalties in the process. If we cannot find a common aim, we shall fall prey to the skeptic's suggestion of an endless proliferation of aims, and the group will splinter; this, of course, is the problem of fragmentation. Why is the finding of a common aim of science such a puzzle? Why cannot one simply say that scientists ought to be interested in the truth? Some philosophers have advocated the notion that science should aim at solving problems; "truth is irrelevant," they have said. That view soon withered away, and for good reasons.[4] Aside from the division among those who hold that the aim of science should be the truth and those who hold that the aim of science should be something other than truth – say, problem solving – there are other deep divisions as well. These divisions make the problem of finding a common aim of science particularly pressing.

The problem can be explicated in this way. Realists, antirealists, metaphysical realists, and constructive empiricists, to mention only a few, are all interested in the truth. Each defines the aim of science in terms of truth, to be sure, but each defines it differently. Realists, like Popper, postulate truth *simplicter* as the aim of science; on his view, it is the task or aim of science to capture the truth about both observable as well as nonobservable things, substances, events, states, or processes. By contrast, an antirealist – say, a good old-fashioned logical empiricist – would deem it the aim of science to capture the truth only about observable things, substances, events, states, or processes. To be sure, antirealists typically have different stories to tell about propositions concerning such nonobservable entities: For example, some antirealists claim that such propositions are meaningless, while other antirealists aver that such propositions have a truth value, but a truth value that we cannot in principle know. Clearly, it is the foregoing differences over the aim of science that constitute our problem, namely, the threat of fragmentation. (Emphatically, the problem is not whether to regard truth as indefinable, as Descartes, Frege, Russell, and Moore had thought; or to define truth in terms of correspondence, coherence, pragmatism, and such; or to regard truth as a minimalist or deflationary notion;[5] or whether,

[4] See my "In Defense of Truth."

[5] Chapter 4 of Paul Horwich's book *Truth*, entitled "Methodology and Scientific Realism," accords well with what I am about here.

following Donald Davidson, we should say that while truth is indefinable we can say revealing things about it by tracing its inviolable interconnections with other basic notions, such as belief, desire, cause, and action.[6])

One can see the implication of the foregoing. Scientists who are realists, and with what they take to be the aim of science, would go off in one direction; their antirealist counterparts would go off in another – not to mention the splinter subgroups in each of these groups. This is the problem of fragmentation we have repeatedly encountered in the book, and it is a problem that is an immediate consequence of not having a common group goal.[7] Our solution to this problem of finding a common group goal, a common aim of science, must consist in showing that (*a*) there is, and can be, only a single philosophically defensible goal of science, or (*b*) while allowing for the philosophical differences between realists and antirealists (as well as differences among their respective subgroups), there is a common core. Their joint scientific enterprise can be defined in terms of the pursuit of that common core. Or finally, (*c*) there is no common core, and a separate problem of group rationality will have to be devised for each group of scientists, a group properly delineated in terms of whatever common goal of science it aspires to. The latter solution, while surrendering to the problem of fragmentation, might attempt to show that fragmentation is fairly limited and, consequently, not as objectionable. To fix our ideas, I will assume that the solution is that truth is the common goal or aim of science (letting the reader specify what he or she takes to be the precise notion and definition of truth, if any, in the present context).

Given the solution to the common-aim problem, our *third* problem is to define the basic structure of the society of scientists. Necessarily, the basic structure must be inextricably linked to the notion of truth, a link I will presently specify. The theories of group rationality discussed in this book take it for granted that the basic structure is to be defined in terms of theories, not methods. Since truth is the ultimate goal of the group, and scientific theories are either true or false, it is hardly surprising that these theories of group rationality unquestioningly assume

[6] Donald Davidson, "The Folly of Trying to Define Truth."
[7] For example, see 81, 127, and 135, this volume.

that the basic structure of the society of scientists must be defined in terms of theories.

I prefer a more circuitous route. Consider the following analogy. John Rawls says in *A Theory of Justice* that just as truth is the first virtue of a system of thought, so justice "is the first virtue of social institutions."[8] By way of examples of social institutions, Rawls cites institutions that distribute rights, duties, and what is cooperatively produced; such institutions would be the monogamous family, competitive markets, legally recognized forms of property, principal economic and social arrangements, and a political constitution with an independent judiciary.[9] What an individual moral agent will eventually end up with – rights, duties, and goods and services – will to a large extent be determined by these institutions.[10] If these institutions are just, the individual will have little to complain of by way of justice; if not, he has powerful grounds on which to protest. Consequently, our aim in ensuring just institutions is, ultimately, to ensure that individuals in the society have been dealt with justly. Justice for individuals is directly dependent on just institutions, institutions that constitute the basic structure of society. Yet we allow some leeway, says Rawls. Our pre-analytic judgments about justice being what they are, we tend to be more modest in our claims. We do not say that there is exactly one system of institutions that is just, but rather that there may be several systems of just institutions, each lying within the "permitted range" (presumably, a range provided by the correct theory of justice). Thus, individuals can be dealt with justly in a variety of ways, under different social institutions or basic structures, duly circumscribed.[11]

Just as social institutions or basic structures deeply affect individuals, so methods profoundly affect individual scientists vis-à-vis the course of science; methods do so because they are used by scientists for

[8] Rawls, *A Theory of Justice*, 3.

[9] See Rawls, *A Theory of Justice*, 6; *Political Liberalism*, 257–9; and *Justice as Fairness: A Restatement*, 10–12.

[10] See *Political Liberalism*, 269–71, wherein Rawls delineates the impact that a societal structure has on the interests, plans, and even desires of the individuals inhabiting that structure. One way of reading the skeptic's concerns about a society of scientists structured according to the obejctivist, one-method view is that it would have deleterious consequences not only on science, but also on the general well-being of the scientists themselves, on their interests, plans, and desires.

[11] Rawls, *A Theory of Justice*, especially 174, 176.

evaluating scientific theories. To explain more fully: A method has two chief components, the objective component and the heuristic component. The objective component defines key terms, such as *paradigm, corroboration,* and *theoretical model.* The heuristic component gives advice about which theories to pursue. The heuristic component may recommend pursuing a theory connected to an installed paradigm; it may recommend overthrowing a research program in a degenerating phase, and so on.[12] Consequently, methods enable scientists to decide which theories to test, develop, experiment with, research, or reject; unsurprisingly, methods have an incalculable impact upon the content, course, and schedule of the history of science. Depending upon the trajectory of that history, scientists' interests are differently impacted: One trajectory (say, immediately after Linneaus or Darwin) may profoundly affect botanists, when another trajectory might have left them indifferent. Once again, we are led to theories and thence to truth (if our theories are true), the ultimate goal of the group; but this time we reach the goal of truth not directly through scientific theories, but through methods as intermediaries that define the basic structure of a society of scientists.

With Rawls, we should also say that adequate theories of group rationality must allow for some legroom. Our pre-analytic judgments about what makes a group of scientists rational are even more tenuous than our judgments about what makes a society just; thus, here we have even more reason to be cautious in our claims. We should not say that there is exactly one system of methods that makes a society of scientists rational, but rather that there may be several systems of methods, each lying within the "permitted range," where the range is defined by the correct theory of group rationality. Thus, the rationality of individual scientists can be preserved under a variety of schemes of methods, where each scheme is duly constrained.

The analogy must not be misunderstood in the following way. Those interested in science studies, for example, might argue that structure should be understood in terms of those institutions – social, religious, moral, political, and economic – that profoundly affect the work of scientists. These undeniably affect the psyche of the scientists in a way in

[12] Since the two components are extensively discussed in the first as well as the final chapter of *A Theory of Method,* I am as brief here as my purpose allows me to be.

which almost nothing else does: The nineteenth-century British entomologist John Obadiah Westwood can hardly be understood in the same way as his no less distinguished twentieth-century counterpart, E. O. Wilson. The science studies may well be genuine empirical studies (for all I know), but I am concerned after matters normative. Thus, taking method as that which defines the structure of a group of scientists is more appropriate – normatively relevant, that is – than taking social, religious, moral, political, or economic structures as defining it. As we saw earlier in discussing the second version of the subjectivist view, the latter way of understanding the relationship between institutions and scientists leads to intractable problems.[13]

The basic structure of a society of scientists, as I have so far argued, is also fundamental in a different way. Rawls had suggested that the basic structure of a society – where the basic structure is defined in terms of the political constitution, the fundamental social and economic arrangements, the structure of the family, and so on – is the primary subject of justice. The principles governing the justice of individual acts are to be understood only against the backdrop of the principles governing the justice of the basic structure of society. Here is how Rawls put it: "Thus the principles for the basic structure of society are to be agreed to first, principles for individuals next, followed by those for the law of nations. Last of all the priority rules are adopted, although we may tentatively choose these earlier contingent on subsequent revision."[14] The analogy can be explored in two ways. First, we can claim that the principles governing the rationality of an individual scientist's decision (to accept or reject a theory) is to be understood only against the backdrop of the principles governing the rationality of the basic structure of a society of scientists. This is an important result, if correct: It shows that group rationality precedes and determines individual rationality (if not wholly, then in significant part).

Here are some examples of natural duty: duty to offer mutual aid, duty to keep promises, duty of reparation, duty of self-improvement, and duty to show respect are examples of positive natural duty; duty of refraining from inflicting injury and duty of not harming the innocent

[13] See Chapter 6, section III, this volume.
[14] Rawls, *A Theory of Justice*, 93.

(duty of nonmaleficence) are examples of negative natural duty.[15] It would appear that these duties can be defined independently of principles defining the just basic structure; they are duties and obligations that transcend those defined by the basic structure.[16] But – and this is the crucial point – even a full and complete list of positive and negative natural duties would not by itself entail principles of social justice. That, indeed, is a separate task.

In a similar fashion, one might argue that even a complete list of the principles that govern individual rationality would not entail principles of group rationality. It is scarcely obvious what would be the parallel to natural duty in the context of individual rationality. If there isn't any parallel, then individual rationality has no role to play that is context-free of group rationality. If there is a parallel – if not, what else was the concern in Peirce's puzzle? – it would show that even a complete set of principles of individual rationality would not constitute or entail the principles of group rationality. In other words, not even an adequate solution to Peirce's puzzle would amount to a solution to the problem of group rationality.[17]

Just as natural duties are utterly crucial in order for the artificial virtues, like justice, to flourish and for a just society to be viable, so also individual rationality may be indispensable if the group is to flourish as well from a rational point of view. For example, suppose that an adequate theory of group rationality is one that proposes that a rational society of scientists is one that is divided into various subgroups, each subgroup being structured along the lines of a distinct method (duly circumscribed). For such a society to flourish, each individual scientist would have to faithfully follow the principles of individual rationality governing his respective subgroup (prescribed by the adequate theory). Notice that even this example takes group rationality as the backdrop, leaving entirely open the question of what would be an example of a principle of individual rationality that is distinct from

[15] Some of these examples of duties are from W. D. Ross's *The Right and the Good*; Ross deemed them prima facie duties, not actual duties. These examples are presented merely for illustrative purposes.

[16] Nevertheless, not even all natural duties – for example, the natural duty to support just institutions – can be defined independently of the just basic structure; see *A Theory of Justice*, 93; see 93–5 for details relevant to this paragraph.

[17] This should reinforce the claims made in Chapter 1, section III.

a principle of individual rationality that follows from, or is entailed by, a principle or theory of group rationality.

There is a second way of charting the analogy. We know, of course, that the several agreements to which Rawls adverts will be collectively arrived at by moral agents in the original position. Similarly, in solving the problem of group rationality, we must show that the following agreements will be collectively arrived at by the scientists (under conditions to be specified, to be sure). First Agreement: It must be agreed that the structure of a group of scientists is to be defined in terms of method.[18] Second Agreement: It must be agreed what the principles of rationality governing the structure of the group are. Third Agreement: It must be agreed what the principles governing individual rationality are, given the second agreement (a point already made). Fourth Agreement: It must be agreed what principles of rationality will govern multiple societies of scientists. Fifth Agreement: It must be agreed what the priority principles are that will enable scientists to decide what to do when principles of group rationality among themselves, or when those principles and principles of individual rationality, conflict. (The subject of the Fourth and Fifth Agreements is, as far as I know, entirely virgin land.)

II. The Council, Reasoning, and Allegiance

Given the crude sketch of the problem, the common aim of science, and the basic structure of a group of scientists, our *fourth* problem is to construct something akin to the Rawlsian notion of the original position, which includes the idea of the veil of ignorance. Let us call it the Council of Scientists. The scientists are to be placed in this Council in which they will collectively reason, given their common aim of science (something they might determine in the Council itself), and arrive at a covenant about how to structure their society, a structure that is defined in terms of method (something, too, they might determine in the Council).

[18] It is assumed, of course, that agreement over the precise statement of the problem of group rationality as well as over what is the aim of science (the collective aim) constitutes the necessary backdrop against which the five agreements that follow are to be understood.

Now, Rawls has argued that the original position must be marked by a veil of ignorance. That is, in the original position the moral agents will not know anything about themselves – they will know nothing, for example, about their sex, income, wealth, social status, race and ethnic group, natural gifts and defects, rational plan of life, particular psychological dispositions, or even to what generation they belong – that is arbitrary from a moral point of view. In short, moral agents are to be prohibited from having any knowledge from which they can derive an unfair advantage in the bargaining game in the original position.[19] The veil of ignorance is necessary because moral agents are interested not only in maximum liberty compatible for like liberty for all, but also in how goods and services conjointly or collectively produced are to be distributed. This is barely a sketch, but it is enough for our purposes.

One might argue that while the veil of ignorance may be necessary in the original position, it is entirely unnecessary in the Council of Scientists, since scientists are not interested in how what is produced conjointly or collectively, namely, truth, is distributed. When a truth is discovered, it is the rightful possession of *all* the scientists; there is *no* problem of distribution here. But I shall argue that, if the problem is viewed from a different angle, there is. No scientist can reasonably expect everyone else to labor on problems and theories that interest him; nor will he labor on problems and theories that interest others, but not him. Mutatis mutandis for metaphysics and methods. Consequently, if methods and metaphysics are to have a profound influence on the direction of the history of science – determined by which theories are pursued, which left by the wayside – then full knowledge of the commitments pertaining to theories, methods, and metaphysics of the scientists will be a roadblock to a covenant in the Council of Scientists, just as knowledge of the particularities of their persons would block agreement in the original position. There is, therefore, a nice problem about how the Council of Scientists is to be characterized, not only in terms of what *kind* of reasons must they be given with which to arrive at a covenant (more on this later), but also in terms of what

[19] Rawls, *A Theory of Justice*, especially section 24; Rawls, *Justice as Fairness: A Restatement*, 14–18, 87. For a distinction between a thicker and a thinner veil of ignorance, see Rawls, *Political Liberalism*, 273.

limitations must be placed, and why, on what the scientists can know in the Council.

Rawls has also persuasively argued that constraints must be provided on rival conceptions of justice that are to be presented in the original position. These conceptions of justice have to be general, universal, public, ordering, and final.[20] Thus, by parallel reasoning, I suggest, an adequate theory of group rationality will have to outline the constraints on methods that are to be presented in the Council as means by which to structure a society of scientists. Just as the constraints eliminate certain conceptions of justice, such as egoism in the form of a first-person dictatorship (for failing to be universal), so also we must have constraints in this field (a new task for meta-methodology, perhaps?) that will eliminate certain rival conceptions of group rationality, such as the subjectivist view (for failing to satisfy a plausible principle of individual rationality, say).

Here is by far the more significant problem, a problem that we have repeatedly faced in the book but never before in this light. Once again, the skeptic comes to our aid. The skeptic demands that we proliferate: proliferate theories, proliferate metaphysics, and proliferate methods. By his lights, why not? The scientists in the Council will have to argue, to decide, whether they wish to structure their society along the lines of a single method or several – but highly eclectic – methods or as the skeptic advocates, multiply methods without cause or restraint. As we have already seen, both versions of the subjectivist view, as well as the objectivist view, advocate (at the very least, implicitly) the structuring of the society of scientists along the lines of a single method. The skeptic, however, is at the other extreme; he wants that society to be structured along the lines of endless methods. The view advocated, by implication, in this book is one of multiple methods (where multiplicity is duly restricted, restricted by an adequate theory of group rationality and meta-methodology).

Now, clearly, the problem about which methods to use to structure a society of scientists is tied to the problem of proliferation of scientific theories. The scientists in the Council will have to argue, to decide, whether they wish to pursue a single scientific theory at a time; pursue several – but highly select – theories simultaneously; or go along with

[20] Rawls, *A Theory of Justice*, 112–18.

the skeptic and pursue an endless multiplication of theories. And since a method enables a scientist to choose a theory for further exploration and testing, the scientists in the Council will inevitably ask, "Which structure will best enable us to significantly fulfill the demands of the principle of proliferation?" I take it as given that the principle of proliferation, classical and widely accepted, is implausible in its unrestricted, skeptical version – for reasons we have already argued. The interesting problem is whether either version of the subjectivist view or the objectivist view – which advocate structuring of a society of scientists along the lines of a single method – will satisfy the more plausible version of the principle of proliferation. Independently of the arguments already presented, it seems fairly unlikely.

Hardly insignificant is the following disanalogy. Let us distinguish, with Rawls, pure procedural justice, perfect procedural justice, and imperfect procedural justice.[21] Basically, what distinguishes imperfect from perfect procedural justice is that in the latter there is an outcome known to be just independently of any procedure, and, furthermore, there is a procedure that guarantees the just outcome. For example, a just distribution of a cake between two individuals is that each should get half, other things being equal, and that there is at least one guaranteed procedure to bring this about. In the case of imperfect procedural justice, on the other hand, while the outcome again is known to be just independently of any procedure, there is no guaranteed procedure for arriving at that outcome. For instance, a legal system should find the innocent innocent and the guilty guilty, but no legal system can guarantee that outcome, though it attempts to produce just that.

Now, to illustrate pure procedural justice. In this case, there is no outcome that is known to be just independently of any procedure. In gambling, for example, provided that the procedure is fair, whatever outcome results (whoever turn out to be the winners and the losers) is *ex hypothesi* just. Rawls had argued that given the rationality accorded to the parties in the original position[22] – say, that the parties are governed by the maximin criterion (maximize your minimum loss), among other things – whatever agreement said parties arrive at

[21] Ibid., 74–5.
[22] Ibid., especially sections 25 and 26.

in that position, under duly restricted or well-defined circumstances, would be deemed just; there is no state they are expected to reach, via an agreement or a covenant, that is just independently of that procedure.

Let us, too, distinguish imperfect from perfect procedural rationality. In the latter, there is an outcome – truth – known independently of any rational procedure, and, furthermore, there is a rational procedure that guarantees the outcome. For example, there might be a method (Bacon, Descartes, and Mill each thought they had such a method) that, if diligently followed, would inevitably produce the right result, namely, truth (where that result is defined independently of the method or procedure). No one today espouses such an optimistic view; every methodologist is now a fallibilist. In the case of imperfect procedural rationality, on the other hand, while the result, truth, is determined independently of any rational procedure, there is no guaranteed procedure, only a probability that the outcome will be achieved. No methodologist would deny this modest claim; he would only claim that his method assures a higher probability of achieving the outcome, truth, than any other extant method.

What of pure procedural rationality in this context? One would have to demonstrate (*A*) that there is no outcome that is known to be rational independently of any procedure, and (*B*) that whatever solution to the problem of group rationality is arrived at in the Council by rational scientists is to be deemed the right or rational solution, and that there is nothing more to be said on the subject. In short, there is no right or rational state a society of scientists is expected to reach, via a covenant among scientists in the Council, that is right or rational independently of that covenant. That, I claim, needs to be shown, not just blandly assumed.

Here's why. If truth or verisimilitude (nearness to the truth) is what the society of scientists is after, then there *is* a state of affairs the scientists wish to reach, the common aim of science, a state that can be defined independently of how the society gets there, independently of any procedure the scientists engage or adopt in the Council, and independently of the basic structure of the society. Therefore, (*A*) is false. Methods must be viewed as a procedure by which the society of scientists attempts to arrive, via theories, at the truth. Evidently, there is no method that can guarantee that truth or verisimilitude will be reached: Just as there is no perfect procedural justice that would serve in the original position, so also there is no perfect procedural rationality that will serve the scientists in the Council. At best, there is only probability. Thus, in the context of group rationality, the Council of Scientists will try to arrive at that structure – structure defined in terms of methods – that will give the society of scientists a higher probability of reaching the truth than any other structure.[23] Now, since the end state, the desired outcome – truth – can clearly be defined independently of any procedure the scientists adopt, it is at best fanciful to think along the lines of pure procedural rationality. There simply is no argument that will show that *whatever* the agreements reached in the Council they will naturally, or probably, lead to the truth; nothing short of coincidence will do that.[24] Therefore, (*B*) is false, too. The falsity of (*A*) entails the falsity of (*B*). I conclude: Pure procedural rationality is neither necessary nor sufficient for what the scientists are about in the Council. At the very least, they will need imperfect procedural rationality.

This brings us down to one of the deepest levels in our discussion and to our *fifth* problem, the problem that so interested Hilary Putnam and that should interest us. Putnam focused on the problem of individual rationality. Here I am merely attempting to demonstrate how individual rationality and group rationality are connected, not offering any theory of that specific connection. Granting, then, that the common aim of the scientists in the group is truth; that the structure of the group must be defined in terms of methods; that the society should be structured along the lines of plausible, multiple methods; and that the society should pursue a multiplicity of scientific theories in accordance with the principle of proliferation, our

[23] It is well known that in Rawls's theory of justice the parties in the original position have no information, particular or general, on the basis of which probabilities can be calculated, and which would then skew their decisions; see Rawls, *A Theory of Justice*, 134, 149–50. Now, shouldn't the scientists in the Council be given information on the basis of which they can calculate how probable their success would be if they structured their society in a certain way? Will access to that knowledge, then, provide them with the means by which they can secure unfair scientific advantages for themselves?

[24] To presuppose that there is an argument that provides a rational link between the covenant of the scientists and the end state, truth, is already to abandon pure procedural rationality.

fifth problem is to carefully delineate the rationality of the parties in the Council. It must be a kind of rationality that will not lead scientists to an impasse – "the no-agreement point," as Rawls would say[25] – and that will lead to a universally acknowledged theory of group rationality by which they will rationally agree to structure their society of scientists.

What sort of rationality, then, shall we accord the scientists in order to enable them to arrive at the aforementioned agreements; in other words, how, on what basis, and for what reasons, will the scientists reach a universally acknowledged solution to the problem of group rationality? The *nature* of reasoning that is involved in the Council of Scientists is worth our attention. The scientists are trying to decide what the structure of their society should be – should it be a society founded on no method, on a single method, on several methods, or with no limit on multiplicity? Let us call this the *form solution* and distinguish it from the *content solution*. After the Council has arrived at the form solution, it will arrive at the content solution, a solution that will determine what *specific* methods they must use to structure their society. Thus, the sort of reasoning the scientists will use to arrive at the aforementioned agreements will be *general* reasoning; that is, reasoning not tied to airing and evaluating the virtues and vices of particular methods. (It is worth asking whether a substantive, yet general, solution to the problem of group rationality – the form solution – can be arrived at solely on the basis of practical rationality, as one might claim can be done in ethics.)

Drawing on the distinction that drove much of the discussion in this book, the distinction between individual rationality and group rationality, we might say that just as principles of justice that apply to individuals are markedly distinct from principles that apply to institutions, so also principles of individual rationality that apply to individual scientists are to be clearly distinguished from principles that govern group rationality (even if the former are dependent on the latter). On the one hand, when the rationality of an individual scientist is in question, we are asking whether a scientist's decision (to accept or reject a theory, for instance) is rational. On the other hand,

[25] Rawls, *A Theory of Justice*, 118.

when the rationality of a group structure is in question, we are asking whether the method that defines the structure of the group is rational. *Rational* is an ambiguous term. In the latter case, it is applied to structure; in the former case, it is not. Failure to draw that distinction, not only causes confusion, it also seriously distorts the problem of group rationality and its solution. Ultimately, it also undermines the significance of two enormously important problems and how they are interrelated.

And that will lead us to our *sixth* problem. It is a task centered on the question "What, if anything, will compel or command a scientist to maintain allegiance to his group or subgroup?" This task spotlights the individual scientist and, as it relates to him, raises numerous questions, such as these: *Qua* scientist, what *am* I? *Qua* scientist, what should I *be*? *Qua* scientist, why should I pursue truth or verisimilitude? *Qua* scientist, what do I owe fellow scientists? *Qua* scientist, why should my allegiance lie to the group (or to my subgroup)? *Qua* scientist, what do I owe the scientists who will come after me? *What* will sustain my allegiance? (Perhaps none of these questions can be answered without invoking the bedrock Putnamian idea.)

It is a fair speculation that the *kind* of society in which a scientist will flourish will be a society that is as mindful of his or her scientific pursuits – given the agreements that have been arrived at in the Council, in which the scientist was a willing participant and has rationally consented to agreements regarding its basic structure – as it is of the pursuits of other scientists, thus producing a group in which it is in the interest of each scientist to aid in the cause of sustaining that society or group. Such a society would ensure, as Rawls would say, a stable society. Such a society would be an indispensable backdrop against which the scientists may entertain, and answer, however tentatively, the questions posed in the previous paragraph – answers that the nature of the society of scientists itself will beckon each scientist to give, answers that will engender group unity and cohesiveness. In short, a rational society of scientists will promote stability, while both its rationality and stability will engender in the scientists a desire to do what they can to sustain their society; it will be their way of showing allegiance. The scientists possess a sense of rationality (as moral agents possess a sense of justice), and the scientists' showing of allegiance is the expression of their "common understanding," born of that sense of rationality, of what it is for a society of scientists to be rational.[26]

III. The Universal Law of Rationality, the Worth of Science, and Utopias

Might there be a single principle of rationality that ultimately governs the scientists in the group? Call this our *seventh* problem. Let us say – it is an old Kant-like point – that there is an autonomous domain of the moral (just as there is an autonomous domain of the beautiful). What makes an action moral is that it is done for the *sake* of morality, and not for the sake of happiness, or perfection, or for any of myriad other reasons. This view comes attached to a metaphysical view of man – man as *Homo noumenon,* and not merely as *Homo phaenomenon.*[27] Even animals can have reason, as Hume thought, but their kind of reason conditions them to act in deterministic ways, just as their desires condition them to act in specific ways. Thus, animals have no free choice. By contrast, man is accorded will and reason that are autonomous; he freely chooses to act in accordance with what his kind of reason dictates. The reason of man self-legislates the Universal Law, the Categorical Imperative – *I ought never to act except in such a way that I could also will that my maxim should become a universal law*[28] – and if every human being were to act in accordance with that law, then that would usher into existence a Kingdom of Ends, a Social Utopia worthy of man.[29]

[26] For the relationship between a just society, its stability, and the role moral agents play in sustaining that stability, in Rawls's argument, see *A Theory of Justice,* especially, 49, 119, 214, 400–1.

[27] Immanuel Kant, *The Metaphysics of Morals,* 6:418. Of course, it was a doctrine central to the *Critique of Pure Reason;* see, for example, A444/B72–A451/B479 and A532/B560–A558/B586.

[28] Immanuel Kant, *Groundwork of the Metaphysics of Morals,* 4:402.

[29] "The concept of every rational being as one who must regard himself as giving universal law through all the maxims of his will, so as to appraise himself and his actions from this point of view, leads to a very fruitful concept dependent upon it, namely that *of a kingdom of ends.*" Then, a few pages later: "A kingdom of ends is thus possible.... Now, such a kingdom of ends would actually come into existence through maxims whose rule the categorical imperative prescribes to all rational beings *if they were universally followed.*" Kant, *Groundwork of the Metaphysics of Morals,* 4:433 and 4:438, respectively.

Let me attempt a Kantian parallel. Say that there is an autonomous domain of the scientific (just as there is an autonomous domain of the moral or of the beautiful). What makes a decision to accept or reject a scientific theory rational, for example, is that it is done for the *sake* of science, and not for the sake of happiness, or social amelioration, or for any other non-science-related reason.[30] Even this view might come wedded to a metaphysical view of man.[31] Man, said Aristotle, has an innate desire to know; the fulfilling of that desire might produce a state toward which he strives, and in that striving he is guided by reason. Why should that state be of intrinsic good or intrinsic value? – *that* is the cardinal question. Man freely chooses to act, insofar as his science is concerned, in accordance with what his reason dictates. Let us say that there is an overarching rational principle that governs his activity as a scientist, and that principle is his Universal Law of Rationality: *I ought*

[30] It would be interesting to speculate what impact this would have on some of the formulations of the problem of group rationality, especially UGR: the Utilitarian Problem of Group Rationality; see Chapter 3, 61.

[31] Both in *A Theory of Justice* as well as in "Justice as Fairness: Political not Metaphysical" and *Political Liberalism*, Rawls offered a theory of justice for a modern democratic society and outlined a principle of tolerance. As he himself acknowledged, however, the metaphysics underlying that theory is conspicuous by its absence. It is Rawls's view that such a metaphysics is not needed. Richard Rorty offers a sharp evaluation of some of these issues in "The Priority of Democracy in Philosophy." I am attempting to show (contrary to both Rawls and Rorty) how important these metaphysical questions are, and why answers to them may well be necessary if, ultimately, a theory of group rationality is to prove adequate.

Rorty says, "For purposes of social theory, we can put aside such topics as an ahistorical human nature, the nature of selfhood, the motive of moral behavior, and the meaning of human life. We treat these as irrelevant to politics as Jefferson thought questions about the Trinity and about transubstantiation." "The Priority of Democracy in Philosophy," 180. And later he asserts, by way of a liberal response, "even if the typical character types of liberal democracies are bland, calculating, petty, and unheroic, the prevalence of such people may be a reasonable price to pay for political freedom" (190). Not only Nietzsche would have wept at hearing this.

Perhaps a Rortyite philosopher of science might say something similar in the case of group rationality, parroting what Rorty says in the case of political theory or social justice. He might say: "Let us put aside issues such as that of an ahistorical human rationality, the motive of scientific behavior, and the meaning life takes on when it is devoted to the pursuit of truth. Let us treat these as irrelevant to science and scientists. And then" – so he might continue – "we might end up with a rational group of scientists that is bland, calculating, petty, and unheroic, to be sure; but that is the price we will have to pay to get scientists to form a rational group, give it their allegiance, and advance the cause of science." Such a view is not mine. Indeed, it is in dread of any such view that I have ventured to argue that the metaphysical foundation of a scientific group might show us a way out of Rorty's desultory, dowdy view.

never to act except in such a way that I could also will that my methodology should become a universal law in the practice of science.[32]

In the case of morality, when a man decides to act, he does so while employing a maxim, a subjective principle. Kant claims that the act is moral provided that the maxim can be universalized following the Categorical Imperative, and not otherwise. Arguably, all the various maxims, $m_1, m_2, m_3, \ldots, m_n$, upon which a moral agent might act can be consistently held together and can be deemed to be unified by the Categorical Imperative. In a very loose sense of the term *flow*, the maxims flow from the Categorical Imperative. Our seventh problem is whether there is a deep fundamental law, a Universal Law of Rationality, akin to the Categorical Imperative. If so, we might claim that when a scientist decides to accept or reject a scientific theory for testing, experimenting, researching, or otherwise exploring, he does so on the basis of a variety of methodological rules, $r_1, r_2, r_3, \ldots, r_n$. It is a conjecture upon which this problem is based that all the various methodological rules that guide a scientist in his multifarious decisions can be consistently held together and that they, too, can be unified by, and flow from, the Universal Law of Rationality.

The Universal Law of Rationality might be a principle on which every practicing scientist can agree; this principle might be a force that serves to unify disparate subgroups of scientists, giving character,

[32] Any attempt to read this strictly along Kantian lines would be demonstrably mistaken. Kant aimed to eliminate maxims on which a moral agent acted, or considered acting, if universalizing the maxim results in a contradiction in conception or contradiction in will; see Kant, *Groundwork of the Metaphysics of Morals*, 4:424. Such a maxim would lack moral worth. Nothing remotely resembling that can be shown in the present context.

There is only this much of an analogy: A scientist committed to universalizing the methodology he uses must also give it up if it can be shown to be inadequate. Such a method would lack rational worth around which a (sub)group could organize or structure itself. The utilitarian doctrine (*U*) is to act in such a manner as to produce the greatest happiness of the greatest number; the Kantian doctrine (*K*) is to act on the maxim that one can at the same time will to be a universal law of nature. (*U*) and (*K*) are competing for the apex position in moral reasoning. In a similar fashion, the five theories of group rationality we have studied thus far are each offering an ultimate principle for governing group rationality. What I am urging is that we look *still* deeper to see on what metaphysical bedrock these theories rest. And, if we look, we might emerge with general principles of rationality or metaphysical pictures (or, with luck, both) that serve to (*a*) anchor the Putnamian idea and (*b*) underwrite the covenant of the scientists in a group.

unity, or cohesiveness to the subgroups as well as to the group as a whole. Perhaps it might prove to be the arch principle espoused in the Council of Scientists, a principle that could straddle, structure, and filter their reasoning. It could enable the scientists to overcome the fragmentation that results from Feyerabend's skeptical view, both versions of the subjectivist's view, and (on at least one reading) Putnam's view. Thus, if every scientist were to act in conformity with the Universal Law of Rationality, that could bring to fruition, as Kant might say, a Scientific Utopia, a utopia worthy of the scientists.[33]

I have no argument to support the conjecture that the Universal Law of Rationality (whatever its form and content) exists, waiting to be discovered; I am merely enamored by the beauty and elegance of this possibility. However, if the conjecture proves to be false, no serious harm will have been done: We will then perforce have to come up with a medley of methodological rules, governing both individual and group rationality, and try to justify them as best we can, throwing into the bargain a set of priority rules.

Let us consider this matter differently. When its role as a provider of social good is subtracted, what, if anything – our *eighth* problem – is left of the worth of science? In short, what, if anything, is science ultimately *for*? Can the significance of science be tied to the essential nature of man? If so – and I believe this is the case – then that significance will instruct us, as nothing else will, on what should be our Scientific Utopia.

Science's impact on society is so pervasive, not only on its surface but on its lowest platform and in-between rungs as well, and its impact is of such force, that it is hard to see anything else. Thus we are admonished: build computers that use molecules to perform digital calculations – they have incredible capacities; attempt to reverse the laws of optics – this will make better cell phone antennas; understand how plants communicate – this will enable us to breed agricultural crops more efficiently; create a Global Positioning System – it will

[33] This is advertisement, not argument. It is worth scrutinizing, though, since it is the first attempt to show how group solidarity may be understood, if not also sustained. If it is shown to be mistaken, we shall have learned a good deal of what is required to show how groups can be cohesive, how allegiance might be engendered, based on no more than the rationality of individual scientists (what germinates in the Putnamian idea). This might prove to be a significant way of solving the problem of fragmentation.

transform navigation, warfare, and law enforcement; construct hydrogen fuel cells – they will power all kinds of vehicles, not to mention the resulting slowing down of global warming; or, as one last example, attempt to compress liquid hydrogen into liquid metal – such a metal can be used as a superconductor at room temperature. Virtually no discovery in medicine is innocent of social implications; no technological advance can restrict its consequences to the innocuous.

But it is worth our while to bracket those important, ever-intruding social consequences of science, and to ask, "If none of these social implications of science were present, what would science be important *for*?" or "When the social implications of science are accounted for, is there a residue that would account for the ultimate significance of science?"[34] If not, why answer questions like: What is the internal structure of the underground nest of the Florida harvester ant, *Pogonomyrmex badius*? What is the shape of a proton? Where is the dark matter? How much carbon is being processed by a phytoplankton? Can a pulse of light be created that lasts little more than half a femtosecond ($^1/_2 \times 10^{-15}$ seconds)? Are the newly discovered footprints evidence of a new theropod dinosaur that gave rise to all bird lineages, including the famed archaeopteryx? Is that system 2,400 light years away a solar system like ours in the making? And the puzzlement is compounded when we ask the old-fashioned utilitarian question. How could resources invested in answering these questions result in payoffs that are greater than those invested in solving problems in science that bear foreseeable social consequences?[35]

Our *ninth* and final task naturally follows from the preceding, namely, what manner of society will give rise to, cater to, and sustain a Scientific Utopia (as defined by the theory of group rationality universally acknowledged in the Council of Scientists). Now, I have contended that the task of group rationality should raise questions

[34] Putnam puts the issue in this way: "Moreover, we are not – nor were we ever – interested in knowledge *only* for its practical benefits; curiosity is coeval with species itself, and pure knowledge is always, to some extent, and in some areas, a terminal value even for the least curious among us." *Pragmatism: An Open Question*, 73.

[35] The standard answer that science will walk to its discoveries at its own pace – and will neither be hurried nor directed – will not do. For even the number of scientific problems that would have tangible social results and that clamor for funding far exceeds our resources.

about Scientific Utopia, and not about Social Utopia – leaving in suspension for now the question "Could a Scientific Utopia flourish in a society far, far distant from a Social Utopia?" (Answer: I suspect not; this is presumably what is implied in the Williams problem). But the argument I offered earlier helps here: Even if a Social Utopia were in existence, it would not settle the question "What manner of Scientific Utopia should there be?"

Let us end here. No satisfactory theory of group rationality can fail to sketch an answer to these nine problems. Each of the five theories of group rationality we have examined – the skeptic's view, the two versions of the subjectivist's view, the objectivist's view, and Putnam's view – conjectured answers to many of the questions posed here, in significant part prompted by the skeptic's view, and each view fell short in a place or two. We also outlined the Scientific Utopia each theory of group rationality entailed (some utopias appeared not too enticing). What novel vision there is in this book is tucked away in the implications of what it has said about these other theories, the classical theories of group rationality. Presenting that vision – explicitly and in full garb – lies entirely outside the scope of this book; it is left as the task for another season.

Bibliography

Anderson, Frank J. *An Illustrated History of the Herbals.* New York: Columbia University Press, 1977.

Appel, Toby A. *The Cuvier-Geoffroy Debate: French Biology in the Decades before Darwin.* New York: Oxford University Press, 1987.

Axelrod, Robert. *The Complexity of Cooperation: Agent-Based Models of Competition and Collaboration.* Princeton, New Jersey: Princeton University Press, 1997.

The Evolution of Cooperation. New York: Basic Books, 1984.

Ben-Menahem, Yemima, editor. *Hilary Putnam.* New York: Cambridge University Press, 2005.

Bernstein, Richard J. "The Pragmatic Turn: The Entanglement of Fact and Value." In Yemima Ben-Menahem, editor. *Hilary Putnam.* New York: Cambridge University Press, 2005, pp. 251–65.

Bicchieri, Cristina. *Rationality and Coordination.* New York: Cambridge University Press, 1993.

Binmore, Ken. *Game Theory and the Social Contract.* Volume 1: *Playing Fair.* Cambridge, Massachusetts: MIT Press, 1994.

Blackburn, Simon. *Ruling Passions: A Theory of Practical Reasoning.* New York: Oxford University Press, 1998.

Brooke, M. de L., and Davies, N. B. "Egg Mimicry by Cuckoos *Cuculus canorus* in Relation to Discrimination by Hosts." *Nature,* vol. 335 (1988), 630–2.

Browne, Janet. *Charles Darwin: The Power of Place.* Volume 2. Princeton, New Jersey: Princeton University Press, 2002.

Cartwright, Nancy, Cat, Jordi, Fleck, Lola, and Uebel, Thomas E. *Otto Neurath: Philosophy between Science and Politics.* New York: Cambridge University Press, 1996.

Cavell, Stanley. *The Claim of Reason: Wittgenstein, Skepticism, Morality, and Tragedy.* New York: Oxford University Press, 1999.

Churchland, Paul M. "To Transform the Phenomena: Feyerabend, Proliferation, and Recurrent Neural Networks." In John Preston, Gonzalo

Munevar, and David Lamb, editors. *The Worst Enemy of Science? Essays in Memory of Paul Feyerabend.* New York: Oxford University Press, 2000, pp. 148–58.

Colbert, Edwin H. *Dinosaurs: An Illustrated History.* Maplewood, New Jersey: Hammond, 1983.

Conradt, L., and Roper, T. J. "Group Decision-making in Animals." *Nature*, vol. 421 (January 9, 2003), 155–8.

Cronin, Helena. *The Ant and the Peacock.* New York: Cambridge University Press, 1999.

Darwin, Charles. *On the Origin of Species.* A Facsimile of the first edition with an introduction by Ernst Mayr. Cambridge, Massachusetts: Harvard University Press, 1976.

Davidson, Donald. "On the Very Idea of a Conceptual Scheme." In his *Inquiries into Truth and Interpretation.* New York: Oxford University Press, 2001, pp. 183–98.

"Epistemology and Truth." In his *Subjective, Intersubjective, Objective.* New York: Oxford University Press, 2001, pp. 177–91.

"The Folly of Trying to Define Truth." In Simon Blackburn and Keith Simmons, editors, *Truth.* New York: Oxford University Press, 2000, pp. 308–22.

Davis, Morton, D. *Game Theory: A Nontechnical Introduction.* New York: Dover, 1997.

Dawkins, Richard. *Unweaving The Rainbow: Science, Delusion and the Appetite for Wonder.* New York: Houghton Mifflin, 1998.

Climbing Mount Improbable. New York: Norton, 1997.

The Selfish Gene. New Edition. New York: Oxford University Press, 1989.

DeNault, L. K., and MacFarlane, D. A. "Reciprocal Altruism between Male Vampire Bats, *Desmodus rotundus.*" *Animal Behavior*, vol. 49 (1995), 855–6.

Dondo, Mathurin. *The French Faust: Henri de Saint-Simon.* New York: Philosophical Library, 1955.

Drake, Stillman. *Galileo.* New York: Hill and Wang, 1980.

Dupré, John. *The Disorder of Things: Metaphysical Foundations of the Disunity of Science.* Cambridge, Massachusetts: Harvard University Press, 1996.

Dyson, Freeman. "The World on a String." *The New York Review of Books*, vol. 51, no. 8 (May 13, 2004), 16–19.

Earman, John. "Carnap, Kuhn and the Philosophy of Scientific Methodology." In Paul Horwich, editor. *World Changes: Thomas Kuhn and the Nature of Science.* Cambridge, Massachusetts: MIT Press, 1993, pp. 9–36.

Feyerabend, Paul. *Conquest of Abundance: A Tale of Abstraction versus the Richness of Being.* Chicago: University of Chicago Press, 1999.

Knowledge, Science, and Relativism. Philosophical Papers, Volume 3. Edited by John Preston. New York: Cambridge University Press, 1999.

Killing Time: The Autobiography of Paul Feyerabend. Chicago: University of Chicago Press, 1995.

Against Method. Revised Edition. New York: Verso, 1988.

Farewell to Reason. New York: Verso, 1987.

Problems of Empiricism. Philosophical Papers, Volume 2. New York: Cambridge University Press, 1981.

Realism, Rationalism, and Scientific Method. Philosophical Papers, Volume 1. New York: Cambridge University Press, 1981.

Science in a Free Society. London: N L B, 1978.

Against Method. Thetford, Norfolk: Verso, 1978.

"Logic, Literacy, and Professor Gellner." *The British Journal for the Philosophy of Science*, vol. 27, no. 4 (December 1976), 381–91.

Against Method. Atlantic Highlands: Humanities Press, 1975.

"Against Method: Outline of an Anarchistic Theory of Knowledge." In Michael Radner and Stephen Winokur, editors. *Analyses of Theories and Methods of Physics and Psychology*. Minneapolis: University of Minnesota Press, 1970, pp. 17–130.

Fisher, R. A. *The Genetical Theory of Natural Selection*. New York: Dover, 1958.

Frangsmyr, Tore, editor. *Linnaeus: The Man and His Work*. Berkeley: University of California Press, 1983.

Friedman, Michael. "Remarks on the History of Science and the History of Philosophy." in Paul Horwich, editor. *World Changes: Thomas Kuhn and the Nature of Science*. Cambridge, Massachusetts: MIT Press, 1993, pp. 37–54.

Gould, Stephen Jay. "Darwin's Untimely Burial." In his *Ever Since Darwin: Reflections in Natural History*. New York: Norton, 1977, pp. 39–45.

Hacking, Ian. "Paul Feyerabend, Humanist." *Common Knowledge*, vol. 3, no. 2 (Fall 1994), 23–8.

Hare, R. M. *The Language of Morals*. New York: Oxford University Press, 1952.

Hawthorn, Geoffrey. "Introduction." In Geoffrey Hawthorn, editor. *The Standard of Living*. New York: Cambridge University Press, 1988, pp. vii–xiv.

Hayek, F. H. A. *The Counter-Revolution of Science: Studies on the Abuse of Reason*. Glencoe, Illinois: The Free Press, 1952.

Heaney, Seamus. *Opened Ground: Selected Poems 1966–1996*. New York: Farrar, Straus, and Giroux, 1998.

Hempel, Carl. "Valuation and Objectivity in Science." In R. S. Cohen and L. Laudan, editors. *Physics, Philosophy and Psychoanalysis: Essays in Honor of Adolf Grunbaum*. Boston: D. Reidel, 1983, pp. 73–100.

"Scientific Rationality: Analytic vs. Pragmatic Perspectives." In Theodore F. Geraets, editor. *Rationality Today*. Ottawa: University of Ottawa Press, 1979, pp. 46–58.

Hesse, Mary. "Book Review of Thomas S. Kuhn, *The Structure of Scientific Revolutions*, 1962." *ISIS*, vol. 54, part 2, no. 176 (June 1963), 286–7.

Horwich, Paul. *Truth*. Second Edition. New York: Oxford University Press, 1998.

Hume, David. *An Enquiry Concerning the Principles of Morals*. Edited by Tom L. Beaucham. New York: Oxford University Press, 1998.

Huxley, Thomas H. "Geological Reform." In his *Discourses Biological and Geological: Essays*. New York: Appleton, 1909, pp. 308–42.

Evolution and Ethics and Other Essays. New York: Appleton, 1898.

Kanbur, Ravi. "The Standard of Living: Uncertainty, Inequality, and Opportunity." In Geoffrey Hawthorn, editor. *The Standard of Living*. New York: Cambridge University Press, 1988, pp. 59–69.

Kant, Immanuel. *The Metaphysics of Morals.* Translated and edited by Mary Gregor. New York: Cambridge University Press, 1998.

Critique of Pure Reason. Translated and edited by Paul Guyer and Allen W. Wood. New York: Cambridge University Press, 1997.

Groundwork of the Metaphysics of Morals. Translated and edited by Mary Gregor. New York: Cambridge University Press, 1988.

Kierkegaard, Soren. *The Sickness unto Death: A Christian Psychological Exposition for Upbuilding and Awakening.* Translated and edited by Howard V. Hong and Edna H. Hong. Princeton, New Jersey: Princeton University Press, 1980.

Kitcher, Philip. *The Advancement of Science.* New York: Oxford University Press, 1993.

"The Division of Cognitive Labor." *The Journal of Philosophy,* vol. 87, no. 1 (January 1, 1990), 5–22.

Kropotkin, Petr. *The Essential Kropotkin.* Edited by Emile Capouya and Keitha Tompkins. New York: Liveright, 1975.

Mutual Aid. Boston: Extending Horizons Books, 1914.

Kuhn, Thomas S. *The Road since Structure: Philosophical Essays, 1970–1993 with an Autobiographical Interview.* Edited by James Conant and John Haugeland. Chicago: University of Chicago Press, 2000.

The Structure of Scientific Revolutions. Third Edition. Chicago: University of Chicago Press, 1996.

"Afterwords." In Paul Horwich, editor. *World Changes: Thomas Kuhn and the Nature of Science.* Cambridge, Massachusetts: MIT Press, 1993, pp. 311–41.

"The Trouble with the Historical Philosophy of Science," Robert and Maurine Rothschild Distinguished Lecture (November 19, 1991), Department of the History of Science, Harvard University, 1992.

The Essential Tension. Chicago: University of Chicago Press, 1977.

The Structure of Scientific Revolutions. Second Enlarged Edition. Chicago: University of Chicago Press, 1970.

The Structure of Scientific Revolutions. Chicago: University of Chicago Press, 1962.

Lakatos, Imre. "History of Science and Its Rational Reconstructions." In John Worrall and Gregory Currie, editors. *Philosophical Papers: The Methodology of Scientific Research Programmes.* Volume 1. New York: Cambridge University Press, 1978, pp. 102–38.

"Popper on Demarcation and Induction." In John Worrall and Gregory Currie, editors. *Philosophical Papers: The Methodology of Scientific Research Programmes.* Volume 1. New York: Cambridge University Press, 1978, pp. 139–67.

Lewontin, R. C. *It Ain't Necessarily So: The Dream of the Human Genome and Other Illusions.* New York: New York Review of Books, 2001.

"Doubts about the Human Genome Project." *The New York Review of Books,* vol. 39, no. 10 (May 28, 1992), 31–40.

Margolis, Howard. *Paradigms and Barriers.* Chicago: University of Chicago Press, 1993.

Marx, Karl. *The Eighteenth Brumaire of Louis Bonaparte.* New York: International Publishers, 1963.

Mellor, D. H. "Introduction." In D. H. Mellor, editor. *Philosophical Papers: F. P. Ramsey.* New York: Cambridge University Press, 1990, pp. xi–xxiii.

Mill, John Stuart. *On Liberty.* Edited by Elizabeth Rapaport. Indianapolis: Hackett, 1978.

Miller, Richard W. *Analyzing Marx: Morality, Power and History.* Princeton, New Jersey: Princeton University Press, 1984.

Mivart, St. George Jackson. *On the Genesis of Species.* London: Macmillan, 1871.

Nilsson, Dan, and Pelger, Sussane. "A Pessimistic Estimate of the Time Required for an Eye to Evolve." *Proceedings of the Royal Society of London,* series B, vol. 256 (1994), 53–8.

Nozick, Robert. *Philosophical Explanations.* Cambridge, Massachusetts: Harvard University Press, 1981.

"Persona Grata: An Interview with Robert Nozick." *World Research: Ink,* vol. 1, no. 14 (December 1977), 2–6.

Anarchy, State, and Utopia. New York: Basic Books, 1974.

Nydegger, R. V., and Owen, G. "Two-Person Bargaining, an Experimental Test of the Nash Axioms." *International Journal of Game Theory,* vol. 3 (1974), 239–50.

Polanyi, Michael. *Tacit Dimension.* Magnolia, Massachusetts: Peter Smith Publisher, 1983.

Personal Knowledge: Towards a Post-Critical Philosophy. Chicago: University of Chicago Press, 1974.

Popper, Karl. *Conjectures and Refutations: The Growth of Scientific Knowledge.* Fifth edition (revised). New York: Routledge, 2000.

Realism and the Aim of Science. Totowa, New Jersey: Rowman and Littlefield, 1983.

Objective Knowledge: An Evolutionary Approach. Revised Edition. Oxford: Oxford University Press, 1979.

"Normal Science and Its Dangers." In Imre Lakatos and Alan Musgrave, editors. *Criticism and the Growth of Knowledge.* New York: Cambridge University Press, 1970, pp. 51–8.

The Open Society and Its Enemies. Volumes 1 and 2. London: Routledge and Kegan Paul, 1962.

The Logic of Scientific Discovery. New York: Harper Torchbooks, 1959.

Poundstone, William. *Prisoner's Dilemma: John von Neumann, Game Theory, and the Puzzle of the Bomb.* New York: Anchor Books, 1992.

Putnam, Hilary. *The Collapse of the Fact/Value Dichotomy and Other Essays.* Cambridge, Massachusetts: Harvard University Press, 2002.

Realism and Reason: Philosophical Papers. Volume 3. New York: Cambridge University Press, 2002.

Words and Life. Edited by James Conant. Cambridge, Massachusetts: Harvard University Press, 2000.

Pragmatism: An Open Question. Malden, Massachusetts: Blackwell, 1995.

Renewing Philosophy. Cambridge, Massachusetts: Harvard University Press, 1992.

Reason, Truth and History. New York: Cambridge University Press, 1990.
The Many Faces of Realism. La Salle, Illinois: Open Court, 1987.
"The 'Corroboration' of Theories." In his *Mathematics, Matter, and Method: Philosophical Papers.* Volume 1. New York: Cambridge University Press, 1981, pp. 250–69.
"Retrospective Note (1978): A Critic Replies to His Philosopher." In Ted Honderich and Myles Burnyeat, editors. *Philosophy as It Is.* Harmondsworth: Penguin, 1979, pp. 377–80.
Meaning and the Moral Sciences. London: Routledge and Kegan Paul, 1978.
Putnam, Ruth Anna. "Doing What One Ought to Do." In George Boolos, editor. *Meaning and Method: Essays in Honor of Hilary Putnam.* New York: Cambridge University Press, 1990, pp. 279–93.
Quine, W. V. O. *The Web of Belief.* New York: Random House, 1970.
Ramsey, F. P. "Truth and Probability." In D. H. Mellor, editor. *F. P. Ramsey: Philosophical Papers.* New York: Cambridge University Press, 1990, pp. 52–109.
Rapoport, Anatol. "The Use and Misuse of Game Theory." *Scientific American* (December 1962), 108–18.
Fights, Games, and Debates. Ann Arbor: University of Michigan Press, 1960.
Raup, David M. *The Nemesis Affair: A Story of the Death of Dinosaurs and the Ways of Science.* New York: Norton, 1986.
Rawls, John. *Political Liberalism.* New York: Columbia University Press, 2005.
Justice as Fairness: A Restatement. Edited by Erin Kelly. Cambridge, Massachusetts: Harvard University Press, 2003.
A Theory of Justice. Revised Edition. Cambridge, Massachusetts: Harvard University Press, 1999.
"Justice as Fairness: Political Not Metaphysical." *Philosophy and Public Affairs,* vol. 14 (Summer 1985), 223–51.
"Social Unity and Primary Goods." In Amartya Kumar Sen and Bernard Williams, editors. *Utilitarianism and Beyond.* New York: Cambridge University Press, 1982, pp. 159–85.
Reuger, Alexander. "Risk and Diversification in Theory Choice." *Synthese,* vol. 109 (1996), 263–80.
Ridley, Matt. *The Origins of Virtue.* New York: Penguin, 1998.
Rorty, Richard. "The Priority of Democracy in Philosophy." In his *Objectivity, Relativism, and Truth: Philosophical Papers.* Volume 1. New York: Cambridge University Press, 1991, pp. 175–96.
Rorty, Richard, et al. "Introduction." In Richard Rorty, J. B. Schneewind, and Quentin Skinner, editors. *Philosophy in History: Essays on the Historiography of Philosophy.* New York: Cambridge University Press, 1984, pp. 1–14.
Ross, W. D. *The Right and the Good.* New York: Oxford University Press, 2002.
Rothschild, Emma. *Economic Sentiments: Adam Smith, Condorcet, and the Enlightenment.* Cambridge, Massachusetts: Harvard University Press, 2001.
Rousseau, Jean-Jacques. *On the Social Contract.* Indianapolis: Hackett, 1987.
Royce, Josiah. *The Philosophy of Loyalty.* New York: Hafner, 1971.
Salmon, Wesley C. "Rationality and Objectivity in Science: *or* Tom Kuhn Meets Tom Bayes." In C. Wade Savage, editor. *Scientific Theories.* Minneapolis: University of Minnesota Press, 1990, pp. 175–204.

"Rational Prediction." *The British Journal for the Philosophy of Science*, vol. 32, no. 2 (June 1981), 115–25.

Sarkar, Husain. "Kant: Let Us Compare." *The Review of Metaphysics*, vol. 58 (June 2005), 755–83.

The Toils of Understanding: An Essay on "The Present Age." Macon, Georgia: Mercer University Press, 2000.

"Anti-Realism Against Methodology," *Synthese*, vol. 116, no. 3 (1998), 379–402.

A Theory of Method. Berkeley: University of California Press, 1983.

"In Defense of Truth." *Studies in History and Philosophy of Science*, vol. 14, no. 1 (March 1983), 67–9.

"Popper's Third Requirement for the Growth of Knowledge." *The Southern Journal of Philosophy*, vol. 19, no. 4 (December 1981), 489–97.

"Putnam's Schemata." *The Southwestern Journal of Philosophy*, vol. 10, no. 1 (March 1979), 125–37.

"Against *Against Method*; or, Consolations for the Rationalist." *The Southwestern Journal of Philosophy*, vol. 9, no. 1 (March 1978), 35–44.

Sen, Amartya Kumar. *Rationality and Freedom.* Cambridge, Massachusetts: Harvard University Press, 2002.

Development as Freedom. New York: Anchor Books, 1999.

"Is the Idea of Purely Internal Consistency of Choice Bizarre?" In J. E. J. Altham and Ross Harrison, editors. *World, Mind, and Ethics: Essays on the Ethical Philosophy of Bernard Williams.* New York: Cambridge University Press, 1995, pp. 19–31.

The Standard of Living. Edited by Geoffrey Hawthorn. New York: Cambridge University Press, 1988.

On Ethics and Economics, New York: Basil Blackwell, 1987.

Sen, Amartya Kumar, and Williams, Bernard, editors. *Utilitarianism and Beyond.* New York: Cambridge University Press, 1982.

Shermer, Michael. *In Darwin's Shadow: The Life and Science of Alfred Russel Wallace: A Biographical Study on the Psychology of History.* New York: Oxford University Press, 2002.

Shils, Edward. *Tradition.* Chicago: University of Chicago Press, 1981.

Skinner, Quentin. "Ms. Machiavelli." *The New York Review of Books* (March 14, 1985), 29–30.

Skyrms, Brian. *Evolution of the Social Contract.* New York: Cambridge University Press, 1996.

Smith, Maynard. *Evolution and the Theory of Games.* New York: Cambridge University Press, 1982.

Strogatz, Steven. *SYNC: The Emerging Science of Spontaneous Order.* New York: Hyperion, 2003.

Taylor, Charles. "Atomism." In his *Philosophy and the Human Sciences: Philosophical Papers*, Volume 2. New York: Cambridge University Press, 1985, pp. 187–210.

Taylor, Keith, editor. *Henri Saint-Simon (1760–1825): Selected Writings on Science, Industry, and Social Organization.* New York: Holmes and Meier, 1975.

Thorndike, Lynn. *A History of Magic and Experimental Science.* Volumes 1–8. New York: Columbia University Press, 1958.

Todes, Daniel Philip. *Darwin without Malthus: The Struggle for Existence in Russian Evolutionary Thought.* New York: Oxford University Press, 1989.

"Darwin's Malthusian Metaphor and Russian Evolutionary Thought, 1859–1917." *Isis,* vol. 78 (1987), 537–51.

Walsh, Vivian. "Philosophy and Economics." In J. Eatwell, M. Milgate, and P. Newman, editors. *The New Palgrave: A Dictionary of Economics,* Volume 3. London: Macmillan, 1987, pp. 861–9.

Watkins, John. "Feyerabend among Popperians 1948–1978." In John Preston, Gonzalo Munevar, and David Lamb, editors. *The Worst Enemy of Science? Essays in Memory of Paul Feyerabend.* New York: Oxford University Press, 2000, pp. 47–57.

"Hume, Carnap and Popper." In Imre Lakatos, editor. *The Problems of Inductive Logic.* Amsterdam: North Holland, 1968, pp. 271–82.

Wiggins, David. "Truth, Invention and the Meaning of Life." In his *Needs, Values, and Truth.* Third edition. New York: Oxford University Press, 2002, pp. 87–137.

Wilkinson, Gerald S. "The Social Organization of the Common Vampire Bat I. Pattern and Cause of Association." *Behavioral Ecology and Sociobiology,* vol. 17 (1985), 111–21.

"Reciprocal Food Sharing in the Vampire Bat." *Nature,* vol. 308 (1984), 181–4.

Williams, Bernard. *Morality: An Introduction to Ethics.* New York: Cambridge University Press, 2004.

Truth and Truthfulness: An Essay in Genealogy. Princeton, New Jersey: Princeton University Press, 2002.

Making Sense of Humanity. New York: Cambridge University Press, 1998.

"The Standard of Living: Interests and Capabilities." In Geoffrey Hawthorn, editor. *The Standard of Living.* New York: Cambridge University Press, 1988, pp. 94–102.

Ethics and the Limits of Philosophy. Princeton, New Jersey: Princeton University Press, 1985.

Moral Luck. New York: Cambridge University Press, 1981

Descartes: The Project of Pure Inquiry. New York: Penguin, 1978.

Wittgenstein, Ludwig. *Philosophical Investigations.* Oxford: Basil Blackwell, 1967.

Zahavi, Amotz. "Arabian Babblers: The Quest for Social Status in a Cooperative Breeder." In P. B. Stacey and Walter D. Koenig, editors. *Cooperative Breeding in Birds.* New York: Cambridge University Press, 1990, pp. 103–30.

"The Theory of Signal Selection and Some of Its Implications." In V. P. Delfino, editor. *International Symposium on Biological Evolution* (April 9–14, 1985). Adriatica Editrice, Bari, 1987, pp. 305–27.

"Reliability in Communication Systems and the Evolution of Altruism." In B. Stonehouse and C. Perrins, editors. *Evolutionary Ecology.* London: Macmillan, 1977, pp. 253–9.

Name Index

Subject Index